The Quantum World

The Quantum World

The disturbing theory at the heart of reality

NEW SCIENTIST

New
Scientist

Nicholas Brealey Publishing
Hachette Book Group
53 State Street
Boston, MA 02109
www.nicholasbrealey.com

Also available
as an ebook

Contents

Series introduction

New Scientist's Instant Expert books shine light on the subjects that we all wish we knew more about: topics that challenge, engage enquiring minds and open up a deeper understanding of the world around us. *Instant Expert* books are definitive and accessible entry points for curious readers who want to know how things work and why. Look out for the other titles in the series:

Scheduled for publication in spring 2017:
> *The End of Money*
> *How Your Brain Works*
> *Where the Universe Came From*

Scheduled for publication in autumn 2017:
> *How Evolution Explains Everything about Life*
> *Machines That Think*
> *Why the Universe Exists*
> *Your Conscious Mind*

Contributors

Editor-in-chief: Alison George, *Instant Expert* editor for *New Scientist*.

This book is based on talks at the *New Scientist* masterclass 'The quantum world' and articles previously published in *New Scientist*, together with specially commissioned content.

Academic contributors

Hugo Cable is a senior research associate at the University of Bristol, UK, where he researches quantum computation and quantum sensors. He co-wrote the section 'Noise: the key to quantum technologies' in Chapter 5.

Johnjoe McFadden is a professor of molecular genetics at the University of Surrey, UK. He is one of the pioneers in the emerging field of quantum biology. He wrote the section 'Has life harnessed the power of quantum mechanics?' in Chapter 6.

Kavan Modi is a lecturer at Monash University, Melbourne, Australia, who focuses on quantum information theory. He co-wrote the section 'Noise: the key to quantum technologies' in Chapter 5.

David Tong is a professor of theoretical physics at the University of Cambridge, where he works on quantum field theory and gravity. He wrote the section 'The question of quantum gravity' in Chapter 8.

Vlatko Vedral is a professor of quantum information science at the University of Oxford and the National University of Singapore. He wrote 'How the quantum world was uncovered' in Chapter 1 and sections on quantum computing in Chapter 4.

Thanks also to the following writers and editors:

Peter Aldhous, Gilead Amit, Anil Ananthaswamy, Jacob Aron, Stephen Battersby, Celeste Biever, Michael Brooks, Amanda Gefter, Lisa Grossman, Douglas Heaven, Rowan Hooper, Valerie Jamieson, Richard Webb.

Introduction

'Can nature possibly be so absurd as it seemed to us in these atomic experiments?'

This is the question that physicist Werner Heisenberg discussed late into the night with his mentor Niels Bohr during the 1920s, when they were writing the rulebook for a whole new way of understanding the world.

The quantum world they were uncovering is seriously bizarre: it is a world where things can exist in two places at once and become inexplicably linked, no matter how far apart they are. In the realm of atoms, electron and particles of light, objects seem to change their behaviour when they are being watched. Surely this couldn't be real, pondered Heisenberg.

Today, after almost a century of intense research, we know the answer to Heisenberg's question. In the microscopic world of atoms and their constituents, our common-sense understanding of reality breaks down and different rules apply. Quantum mechanics has never failed an experimental test.

In this *New Scientist Instant Expert* guide we take you on a tour of this weird world and the fascinating characters who uncovered it. They include Albert Einstein, who hated the idea of the 'spooky action at a distance' of quantum mechanics, and Erwin Schrödinger, who devised his famous cat thought experiment to demonstrate the absurdity of this bizarre place.

What does this all mean? Do things only become real when they are observed? Are new universes spawned every time we make a measurement? What does this mean for the bedrock of reality?

As well as these mind-bending questions, quantum mechanics has given us many practical technologies: lasers, nuclear reactors and the transistors that underlie computers and all digital technology. In the future, it promises much more: computers more powerful than any built before, the ability to communicate with absolute privacy, and even quantum teleportation.

This guide also explores the role quantum mechanics plays in biology. Has evolution taken advantage of quantum weirdness in designing the biochemistry of life, from birds' navigation systems to the photosynthesis of plants?

Ideas from quantum mechanics are even beginning to percolate out into the vast scale of the cosmos. Many physicists think that combining it with Einstein's general theory of relativity will reveal a new understanding of the Big Bang and the nature of space and time.

This guide gathers together the thoughts of leading physicists and the best of *New Scientist* magazine to bring you up to date with the past, present and future of the quantum world, its applications and intriguing implications.

Alison George, Editor-in-Chief, *Instant Expert* guides

I
Welcome to the weird

The discovery of the quantum world was kick-started by what its originator called 'an act of desperation' at the end of the nineteenth century. This chapter describes how this new field of physics arose and developed.

How the quantum world was uncovered

As a young student, the German physicist Max Planck (1858–1947) had been told by his university professor that, in physics, 'almost everything is already discovered and all that remains is to fill a few holes'. When, in his forties, Planck (see Figure 1.1) decided to tackle one of these minor problems, in the process he inadvertently gave rise to a revolutionary new field of physics.

The problem that Planck investigated was the radiation released by a black body – a perfect absorber and radiator of energy – that defied explanation by the existing laws of

FIGURE 1.1 Max Planck, the originator of quantum theory, which revolutionized our understanding of atomic and subatomic processes

physics (see 'The laws of classical physics' below). No matter how hot they became, black bodies emitted almost no ultraviolet light.

In 1900 Planck announced his solution to this 'ultraviolet catastrophe': instead of being continuous, energy comes in discrete little packages, which he called quanta. But Planck had no idea why energy should be like this – which is why he called his solution an act of desperation. He had no experimental proof, just a mathematical formula. No one, least of all Planck, appreciated what a radical discovery it was.

That began to change when, five years later, a 25-year-old unknown called Albert Einstein (1879–1955) (see Figure 1.2)

FIGURE 1.2 Albert Einstein in 1904; his work on the photoelectric effect led him to propose the concept of the photon.

proposed an even more revolutionary idea. He was working on the photoelectric effect, the phenomenon whereby electrons are released from metals by certain frequencies of light, regardless of the light's intensity. He argued that, if energy could be transmitted in discrete packets, then so, too, could light. He proposed that light, rather than being a continuous wave, is made up of a stream of little 'atoms' called photons. Although he is best known for his theory of relativity, Einstein called his 1905 paper in which he proposed the concept of a photon his 'only revolutionary one'.

The traditional understanding of light and energy was beginning to fall apart. A further breakthrough came from Einstein's Danish contemporary Niels Bohr (1885–1962), who was struggling with the fact that, according to the classical laws of physics, atoms should not exist. Inside the atom, negatively charged electrons whizz around a positively charged nucleus, but, in theory, these electrons would lose energy and ultimately spiral into the nucleus. The stability of matter was impossible.

Bohr resolved this by proposing that electrons move around in discrete orbits and cannot exist between these two orbits. If they jump between two orbits, they emit photons. His calculations of the frequencies of these photons agreed perfectly with the data from experiments at that time. This was another confirmation that light is emitted in small packets whose energy matches the energy difference between these two levels of electrons.

The laws of classical physics

The English scientist Isaac Newton (1643–1727) imagined the universe to be like a giant clockwork machine running according to his immutable laws of motion, devised in the 1680s. Once the initial conditions have been set, the universe evolves deterministically.

The laws of classical Newtonian physics were subject to many tests during the eighteenth and nineteenth centuries. They provide such an accurate description of events at the macro level, from the motion of billiard balls to the movement of planets, that the Nobel prize-winning physicist Albert Michelson (1852–1931) famously wrote in 1903 that 'the more important fundamental laws and facts of physical science ... are so firmly established that the possibility of their ever being supplanted in consequence of new discoveries is exceedingly remote'.

Quantum physics, however, has changed this picture in the most dramatic way. In quantum physics the notion of chance enters at a fundamental level. When a quantum particle, such as a particle of light – a photon – encounters a piece of glass, such as your window, it seems to behave randomly. There is a chance that it will go through, but there is also a chance that it will be reflected. As far as we can tell, there is nothing in the universe that determines which alternative will happen at any given time.

Then something even more surprising happened, thanks to French physicist and aristocrat Louis de Broglie (1892–1987) (see Figure 1.3). If light waves are also particles, he argued,

FIGURE 1.3 Louis de Broglie, who showed that electron 'particles'
also behave as 'waves'

then why not think about nature in a unified way, and suppose that atoms and electrons are also wavelike? Building on Einstein's photon equations, he showed that electron 'particles' also behave as 'waves'.

Einstein loved this revolutionary idea but, at that time, it was just a hypothesis. De Broglie had no experimental evidence. However, his work gave huge momentum to new research. Soon, experiments on electrons and helium atoms confirmed that they really do behave like waves – they scatter and create interference patterns when they pass through a grating in the same way as water would (see Figure 2.1). Despite defying our common-sense notions, wave–particle duality became accepted as reality.

In 1927 Werner Heisenberg (1901–76), one of Bohr's most brilliant students, realized that one of the consequences of wave–particle duality was that it put a fundamental limit to how much information we could ever know about a physical system. The more precisely we measure a particle's location, the less we can know about its momentum. This uncertainty is nothing to do with the practical challenge of making measurements at the scale of photons and electrons – it is a fundamental feature of the universe. Heisenberg showed that, in the quantum world, objects do not have separate properties of momentum and position as they do in our everyday world. They have a mixture of the two, which can never be completely separated. Even today, Heisenberg's uncertainty principle remains one of quantum theory's most unsettling predictions.

Bohr and Heisenberg

Some middle-aged men have trains in their attics. Niels Bohr had Werner Heisenberg. In the winter of 1926–7, the brilliant young German was working as Bohr's chief assistant, billeted in a garret at the top of the Bohr's Copenhagen institute. After a day's work, Bohr would come up to Heisenberg's eyrie to chew the quantum fat. They often sat up late into the night, in intense debate over the meaning of the new, revolutionary quantum theory.

An equally unsettling proposition came at around the same time from the Austrian physicist Erwin Schrödinger (1887–1961). In 1925, after giving a talk, he was asked by one of the audience: 'You keep talking about electrons and atoms being waves but shouldn't they obey a wave equation? You never mentioned a wave equation in your talk.' That equation did not yet exist. So Schrödinger went away skiing over the weekend and came back with that equation – now known as the Schrödinger equation – describing how a quantum system changes with time. The kind of equation to describe the behaviour of water or light could now be applied to atoms and molecules. This led to even more outcomes that defy our everyday notions of how the world works.

Schrödinger agreed that we cannot describe a particle as inhabiting a fixed point in space. Instead, he said that we can only assign a set of possibilities to all the possible positions where it could exist, and a particle only settles into a specific location when someone takes the trouble to look at it.

The absurdity of this logic weighed heavily on the quantum physicists of that time and at this point Einstein became an opponent of the field. Even those who are used to the idea feel uncomfortable about the implications of Schrödinger's equation. It means that a particle can be in two places at the same time and, when measured, it will seem to randomly pop up in only one of them. From any given starting condition, quantum mechanics cannot predict an outcome, unlike in the classical world of physics, leading to Einstein's famous criticism that God doesn't play dice.

Einstein's other major complaint was about the strange quantum phenomenon of entanglement, where two particles are linked, no matter far apart they are. When a photon is sent to a beam splitter called an interferometer, it goes both ways at the same time. If you make a measurement in one arm of the beam splitter and you don't detect a photon, it means you have instantaneously created a photon in the other arm of the interferometer. Even if the two arms were thousands of light years apart, the detection of a photon in one arm would cause a photon to be created or destroyed instantaneously in the other arm, despite the vast distance between them. How could that be? Einstein didn't like this 'spooky action at a distance' because it appeared to violate the laws of relativity, which state that nothing can travel faster than the speed of light. He also complained that quantum mechanics provided no deeper description about why this happened (see 'Faraday's "magic"' below).

Faraday's 'magic'

During the 1820s the pioneer of electromagnetism Michael Faraday (1791–1867) used to perform a trick in his Christmas lectures at the Royal Institution in London that seemed similar to the 'spooky action at a distance' of quantum mechanics.

Faraday had a large coil with a magnet at one end, and at the other end, some distance away, was a compass. When he put the magnet in the coil, the needle on the compass moved, even though it was nowhere near the magnet. To the audience it looked like magic – like spooky action at a distance. In reality, the movement of the compass was caused by the changing magnetic field in the coil. But, at that time, the concept of a magnetic field had not yet been developed.

With quantum mechanics we are also lacking such an explanation. When I make a measurement in one place, things collapse in another location, seemingly randomly and without any reason. As far as we know, there's nothing happening in between the two places – there is no equivalent of electrons moving down Faraday's coil.

Even today, scientists are grappling with the philosophical conundrums raised by quantum mechanics. What do they mean? Is nothing really real until it is observed? The Australian-born

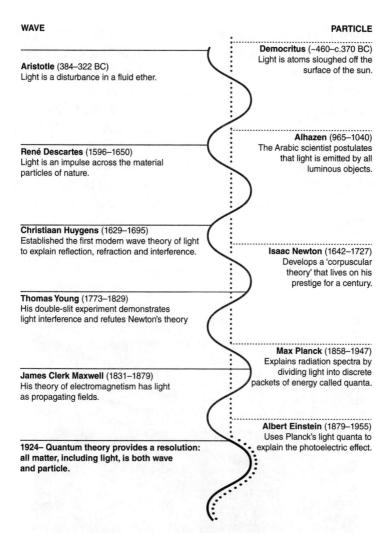

WAVE

PARTICLE

Democritus (~460–c.370 BC)
Light is atoms sloughed off the
surface of the sun.

Aristotle (384–322 BC)
Light is a disturbance in a fluid ether.

Alhazen (965–1040)
The Arabic scientist postulates
that light is emitted by all
luminous objects.

René Descartes (1596–1650)
Light is an impulse across the material
particles of nature.

Christiaan Huygens (1629–1695)
Established the first modern wave theory of light
to explain reflection, refraction and interference.

Isaac Newton (1642–1727)
Develops a 'corpuscular
theory' that lives on his
prestige for a century.

Thomas Young (1773–1829)
His double-slit experiment demonstrates
light interference and refutes Newton's theory

Max Planck (1858–1947)
Explains radiation spectra by
dividing light into discrete
packets of energy called quanta.

James Clerk Maxwell (1831–1879)
His theory of electromagnetism has light
as propagating fields.

Albert Einstein (1879–1955)
Uses Planck's light quanta to
explain the photoelectric effect.

**1924– Quantum theory provides a resolution:
all matter, including light, is both wave
and particle.**

FIGURE 1.4 Duelling over duality: philosophers and physicists have long
argued about whether light is a wave or a particle

British physicist William Lawrence Bragg (1890–1971) even hypothesized that 'everything in the future is a wave, everything in the past is a particle', implying that quantum mechanics may be responsible for the directionality of time.

But the quantum pioneers have been vindicated. Quantum theory has never failed an experimental test. At the subatomic level, the way things are measured really does determine the outcome, and particles and waves are dual aspects of a single reality (see Figure 1.4). Even today, we know of no natural law that prevents quantum mechanics from being true at the level of the universe.

The key players

FIGURE 1.5 The founding figures of quantum mechanics at the Solvay Conference in Brussels in 1927

In October 1927 an incredible meeting of minds took place at the fifth Solvay Conference in Brussels, where the leading physicists of the time met to discuss the new field of quantum mechanics. Of the 29 people in the photograph marking the event (Figure 1.5), 17 were, or would become, Nobel prize-winners (Marie Curie, the only woman in the picture, would even be awarded two). All the founding figures of quantum mechanics were present:

Albert Einstein (front row, fifth from left)

Einstein was only 26 years old when he produced the remarkable series of papers of his *annus mirabilis*, 1905. These included his work on special relativity and the famous equation $E = mc^2$. However, the first major paper he published that year was on the photoelectric effect, marking a major leap forward in the nascent field of quantum mechanics by showing how energy comes in discrete packets. It was this work, and for his 'services to theoretical physics' for which he was awarded the Nobel Prize in 1921. As a Jew, Einstein faced increasing hostility in Nazi Germany and renounced his citizenship in 1933. He eventually found refuge at the Institute of Advanced Study in Princeton, New Jersey, where he remained until his retirement.

Erwin Schrödinger (back row, sixth from left)

When this picture was taken, Schrödinger was still eight years away from formulating his famous cat-centred thought experiment, which exposed the apparent absurdity of quantum mechanics. Born in Vienna, Austria, Schrödinger devised a wave equation to explain the behaviour of quantum systems – work that won him the 1933 Nobel Prize in Physics. An opponent of the Nazi regime, he left Austria in 1934 and eventually went to live in Dublin, Ireland, where he helped to establish an Institute of Advanced Studies. He is also known for his unconventional love life, living with both his wife and his mistress.

Max Planck (front row, second from left)

The grandfather of quantum mechanics, Max Planck was born in Kiel, Germany. Unlike most of the other key players in the field, Planck was relatively old (42) when he proposed that energy comes in discrete packets, but after making this revolutionary discovery, which won him a Nobel Prize in 1918, he played only a minor role in the further development of quantum theory. He remained in Germany, as a professor at the University of Berlin, but his personal life was mired in tragedy. His son Karl was killed in the First World War, both his daughters died in childbirth and the Gestapo executed another of his sons, Erwin, in 1945 due to his suspected involvement in a plot to assassinate Adolf Hitler.

Werner Heisenberg (back row, ninth from left)

Best known for his uncertainty principle, Heisenberg was born in Würzburg, Germany, and after finishing his doctorate he worked for Niels Bohr in Copenhagen. He was awarded the 1932 Nobel Prize in Physics for 'the creation of quantum mechanics' and for his theory of the atom, in which electrons absorb and emit radiation of fixed wavelengths when jumping between fixed orbits around a nucleus. He was also the lead scientist in the Uranverein, Germany's project to develop nuclear technology, and famously met with Bohr in Germany-occupied Copenhagen in 1941 to discuss his dilemmas about this work. After the war he remained in Germany, investigating nuclear energy, cosmic rays and subatomic particles.

Paul Dirac (middle row, fifth from left)

Born in Bristol, UK, Paul Dirac (1902–84) was responsible for a crucial piece of the explanation of fundamental particles and forces. His equation, which he proposed in 1928 to describe an electron travelling at close to the speed of light, married the quantum physics of Schrödinger and Heisenberg with Einstein's special relativity. It also predicted a whole new set of subatomic particles known as 'antiparticles'. He shared the 1933 Nobel Prize in Physics with Schrödinger. He was an eccentric and awkward character: the master of monosyllabic answers, he went through life refusing to engage with colleagues, students and even his own family.

Wolfgang Pauli (back row, eighth from left)

Without the exclusion principle developed by Wolfgang Pauli (1900–58) in 1925, matter as we know it would not exist. His principle says that no two electrons in an atom can enter the same quantum state. This work won him a Nobel Prize in 1945. He also was the first to predict the existence of mysterious neutrino particles in 1930. Soon after this, Pauli had a nervous breakdown, and underwent therapy with the renowned psychoanalyst Carl Jung. He was also a close friend of both Niels Bohr and Werner Heisenberg. Pauli was born in Vienna and, although raised a Catholic, had Jewish roots, which led him to leave for the United States in 1940. After the war he returned to Zurich, where he remained for the rest of his life.

Arthur Compton (middle row, sixth from left)

This American physicist was awarded the Nobel Prize just after the 1927 Solvay Conference 'for the effect named after him', which showed how photons are scattered by charged particles. This was an important piece of the quantum jigsaw puzzle, demonstrating that light cannot be explained as purely a wave phenomenon. Compton (1892–1962) played a key role in the Manhattan Project, the US nuclear weapons programme of the Second World War.

Louis de Broglie (middle row, seventh from left)

Prince Louis-Victor Pierre Raymond de Broglie (to give his full name) was born in Dieppe, France, to a noble family. He wrote an astonishingly original 70-page PhD thesis in 1924, with the title 'Recherches sur la Théorie des Quanta', in which he laid out the principle of wave–particle duality – and was awarded the Nobel Prize in Physics a mere five years later. He presented his 'pilot wave' theory, in which a particle is accompanied by a guiding wave, at the 1927 Solvay Conference, but later abandoned the idea. It was rediscovered in 1952 by the US physicist David Bohm (1917–92), who reformulated it. De Broglie was instrumental in helping to establish CERN, the European Organization for Nuclear Research in Geneva, Switzerland.

Max Born (middle row, eighth from left)

The German physicist and mathematician Max Born (1882–1970) was awarded the 1954 Nobel Prize in Physics, 'for his fundamental research in quantum mechanics, especially for his statistical interpretation of the wavefunction', work he had carried out at the University of Göttingen 30 years earlier. During his tenure there, he also supervised many of the future leading lights in the field, such as Heisenberg and Pauli. When the Nazis came to power in the 1930s, Born (a Jew) was suspended and left for the UK.

Niels Bohr (middle row, far right)

Born in Copenhagen, Denmark, Bohr was the first of the quantum pioneers to truly recognize and confront the philosophical problems posed by the theory. His solutions are still debated today. After Denmark was occupied by Germany during the Second World War, Bohr had a famous meeting with Heisenberg, who had become the head of the German nuclear programme. By 1943, under threat of arrest (like many of the quantum pioneers, he too was Jewish), he fled to Sweden, then the UK, where he joined the British mission to the Manhattan Project. After the war he returned to Denmark. He was awarded the Nobel Prize in Physics in 1922 for 'for his services in the investigation of the structure of atoms and of the radiation emanating from them'.

How can a theory that works so well have such unnatural foundations?

Most theories are built on a solid foundation of first principles – not so quantum theory. Although it was initially inspired by an idea rooted in the real world – that energy came in small packets called quanta – by the time luminaries such as Erwin Schrödinger and Werner Heisenberg had finished its mathematical formulation, the theory had acquired a life of its own.

Gone was any certain correspondence between mathematical variables and physical properties. In their place were abstruse objects such as wave functions, state vectors and matrices, all acting in an unreal mathematical environment called Hilbert space – a higher-dimensional, complex version of normal three-dimensional space.

Bizarrely, though, these abstractions work. Follow a set of mathematical rules laid down by the founders of quantum theory and you can make physical predictions that are confirmed time and time again by experiment. Particles that pop up out of nothingness only to disappear again, objects whose physical states can become 'entangled' and can influence each other instantaneously over vast distances, cats that remain suspended between life and death as long as we don't look at them: all of these flow from the mathematical formulation of quantum theory, and all seem to be true reflections of how the world works.

A brief history of the quantum revolution

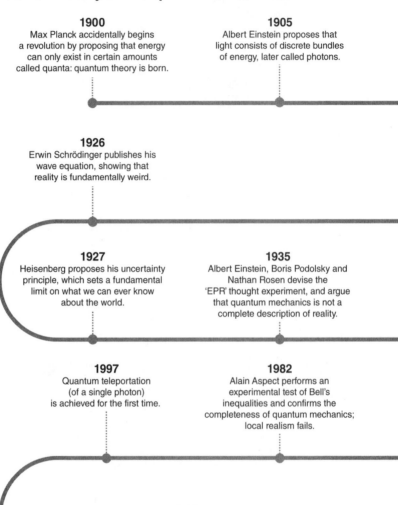

1900
Max Planck accidentally begins
a revolution by proposing that energy
can only exist in certain amounts
called quanta: quantum theory is born.

1905
Albert Einstein proposes that
light consists of discrete bundles
of energy, later called photons.

1926
Erwin Schrödinger publishes his
wave equation, showing that
reality is fundamentally weird.

1927
Heisenberg proposes his uncertainty
principle, which sets a fundamental
limit on what we can ever know
about the world.

1935
Albert Einstein, Boris Podolsky and
Nathan Rosen devise the
'EPR' thought experiment, and argue
that quantum mechanics is not a
complete description of reality.

1997
Quantum teleportation
(of a single photon)
is achieved for the first time.

1982
Alain Aspect performs an
experimental test of Bell's
inequalities and confirms the
completeness of quantum mechanics;
local realism fails.

2015
A loophole-free Bell test confirms
that Einstein was wrong and nature is
indeed quantum mechanical.

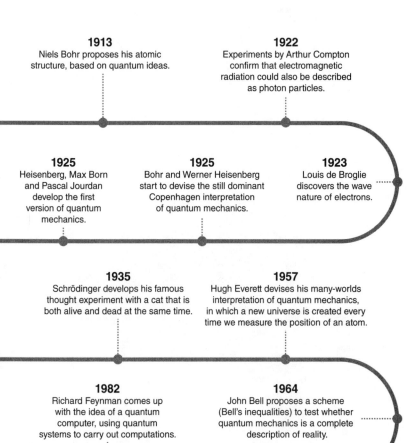

1913
Niels Bohr proposes his atomic structure, based on quantum ideas.

1922
Experiments by Arthur Compton confirm that electromagnetic radiation could also be described as photon particles.

1925
Heisenberg, Max Born and Pascal Jourdan develop the first version of quantum mechanics.

1925
Bohr and Werner Heisenberg start to devise the still dominant Copenhagen interpretation of quantum mechanics.

1923
Louis de Broglie discovers the wave nature of electrons.

1935
Schrödinger develops his famous thought experiment with a cat that is both alive and dead at the same time.

1957
Hugh Everett devises his many-worlds interpretation of quantum mechanics, in which a new universe is created every time we measure the position of an atom.

1982
Richard Feynman comes up with the idea of a quantum computer, using quantum systems to carry out computations.

1964
John Bell proposes a scheme (Bell's inequalities) to test whether quantum mechanics is a complete description of reality.

2016
Quantum teleportation over 6.2 km of cables is achieved.

2
A tour of the quantum world

From undead cats to particles popping up out of nowhere and ghostly influences at a distance, quantum physics delights in demolishing our intuitions about how the world works. Here is a tour of its key properties.

Wave–particle duality

The oldest and grandest of the quantum mysteries relates to a question that has exercised great minds at least since the time of the ancient Greek mathematician Euclid more than two thousand years ago: what is light made of? Throughout history, people have debated the issue (see Figure 1.4).

Isaac Newton thought light was tiny particles. Not all his contemporaries were impressed, and in classic experiments in the early 1800s the polymath Thomas Young showed how a beam of light diffracted, or spread out, as it passed through two narrow slits placed close together, producing an interference pattern on a screen behind it, just as if it were a wave.

So which is it, particle or wave? Quantum theory provided an answer soon after it bowled on to the scene in the early twentieth century. Light is both a particle and a wave – and so, for that matter, is everything else. A single moving particle such as an electron can diffract and interfere with itself as if it were a wave and, believe it or not, an object as large as a car has a secondary wave character as it moves along the road.

That revelation came in a wildly successsful doctoral thesis submitted by the pioneering quantum physicist Louis de Broglie in 1924. He showed that, by describing moving particles as waves, you could explain why they had discrete, quantized energy levels rather than the continuum predicted by classical physics. De Broglie first assumed that this was just a mathematical abstraction, but wave–particle duality seems to be all too real. Young's classic wave interference experiment has been reproduced with electrons and all manner of other particles (see Figure 2.1).

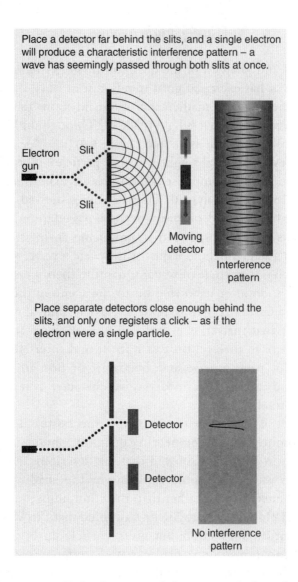

Place a detector far behind the slits, and a single electron will produce a characteristic interference pattern – a wave has seemingly passed through both slits at once.

Electron gun

Slit

Slit

Moving detector

Interference pattern

Place separate detectors close enough behind the slits, and only one registers a click – as if the electron were a single particle.

Detector

Detector

No interference pattern

FIGURE 2.1 Updated versions of Thomas Young's classic double-slit experiment show how particles also look like waves – depending on how you detect them

Einstein versus Bohr

One of the most famous bouts in science was between Albert Einstein and Niels Bohr (see Figure 2.2). For the years spanning the late 1920s and early 1930s, these two fought over the future of physics. Einstein could not accept the outrageous randomness and unknowability of quantum mechanics, so he attacked the theory by devising a series of ingenious thought experiments. But whenever he seemed to have nailed an inconsistency at the core of quantum theory, Bohr proved him wrong. Despite all its unpalatable ingredients, quantum mechanics won the day.

FIGURE 2.2 The Danish physicist Niels Bohr

Entanglement

Entanglement is the idea that particles can be linked in such a way that changing the quantum state of one instantaneously affects the other, even if they are light years apart. This 'spooky action at a distance', in Einstein's words, is a serious blow to our conception of how the world works. Erwin Schrödinger (see Figure 2.3) called it the 'defining trait' of quantum theory. Einstein could not bring himself to believe in it at all, thinking it proof that quantum theory was seriously flawed.

Superposition

No matter how hard you try, you cannot be in two places at once. But if you're an electron, popping up in multiple places is a way of life. The laws of quantum mechanics tell us that

FIGURE 2.3 Erwin Schrödinger

subatomic particles exist in this superposition of states until they are measured and found to be in just one – when their wave function collapses.

So why can't we do the same party trick as an electron? It seems that, once something gets large enough, it loses its quantum properties, a process known as decoherence (see Chapter 7). That is mainly because larger objects interact with their environment, which forces them into one position or another. Erwin Schrödinger famously pointed out the absurdity of large-scale superposition with the example of a cat that is both dead and alive, whose fate depends on the decay of a radioactive atom – a random quantum process.

Schrödinger's wave equation

In 1926 Erwin Schrödinger had put forward the idea that all quantum particles, from atoms to electrons, could be described by intangible entities that spread out through space, much like ripples on a lake's surface. He called them wave functions, and they neatly explained why electrons in atoms have the energies they do.

All waves can be described mathematically – a ripple across a pond, for example, is a disturbance in water; its wave function describes its shape at any point and time, while something called the wave equation (see Figure 2.4) predicts how the ripple moves. Schrödinger realized from de Broglie's work that every quantum system has a wave function associated with it too, although he struggled to explain what the disturbance would be in the case of an atom or an electron. Even so, Schrödinger's work led to a radical new picture of the quantum world as a place in which certainties give way to probabilities.

Schrödinger's wave function is central because it encodes all the possible behaviours for a quantum system. Picture the simple case of an atom flying through space. It is a quantum particle, so you cannot say for sure where it will go. If you know its wave function, however, you can use that to work out the probability of finding the atom at any location you please.

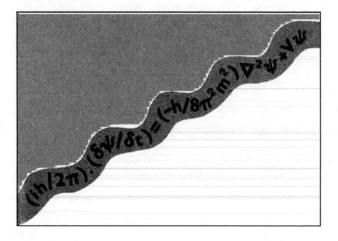

FIGURE 2.4 Schrödinger's wave equation, a basic tool of physics; even now, nobody really knows what it means

Quantization

First Max Planck showed in 1900 that, mathematically, the electromagnetic energy given out by a radiating body was emitted not continuously but in indivisible packets. Five years later Einstein showed that light was made of discrete particle-like quanta that he called photons. This was just the beginning. As quantum theory developed, it became clear that not just energy but many other properties, such as electric charge and spin, come in units of a minimum size. Why that should be no one knows.

Probability

Probabilities are fundamentally different things in classical and quantum physics. In classical physics they are 'subjective' quantities that constantly change as our state of knowledge changes. The probability that a coin toss will result in heads or tails, for instance, jumps from ½ to 1 when we observe the outcome. If there were a being who knew all the positions and momenta of all the particles in the universe – known as a 'Laplace demon' after the French mathematician Pierre-Simon Laplace (1749–1827), who first countenanced the possibility – it would be able to determine the course of all subsequent events in a classical universe, and would have no need for probabilities to describe them.

In quantum physics, however, probabilities arise from a genuine uncertainty about how the world works. States of physical systems in quantum theory are represented in what Schrödinger called catalogues of information, but they are catalogues in which adding information on one page blurs or scrubs it out on another. Knowing the position of a particle more precisely means knowing less well how it is moving, for example. Quantum probabilities are 'objective' in the sense that they cannot be entirely removed by gaining more information.

Spin

Spin is a slippery concept to grasp. A quantum property of many sorts of particles, including electrons, it was first proposed in the early 1920s by the Austrian theoretical physicist Wolfgang Pauli, who was so strong-willed that it was said he could make experiments fail simply by entering their vicinity. With spin, he did not have to. Its signature is seen when you send a stream of electrons through an uneven magnetic field. The particles are deflected in opposite directions seemingly at random, as though each one has an intrinsic rotation that is somehow 'caught' by the magnetic field, causing it to veer off course.

Uncertainty

One puzzle that Bohr and his student Heisenberg (see Figure 2.5) pondered in the winter of 1926–7 was the

FIGURE 2.5 Werner Heisenberg in about 1925

trails of droplets left by electrons as they passed through cloud chambers, an apparatus used to track the movements of charged particles. When Heisenberg tried calculating these seemingly precise trajectories using the equations of quantum mechanics, he failed.

One evening in mid-February, Heisenberg went out for a walk and then it came to him. The electron's track was not precise at all: if you looked closely, it consisted of a series of fuzzy dots. That revealed something fundamental about quantum theory. Heisenberg excitedly wrote his idea down in a letter to fellow physicist Wolfgang Pauli. The gist of it appeared in a paper a few weeks later: 'The more precisely the position is determined, the less precisely the momentum is known in this instant, and vice versa.' Thus Heisenberg's notorious uncertainty principle was born. A statement of the fundamental unknowability of the quantum world, it has stood firm for the best part of a century.

The profound implications of the uncertainty principle are hard to overstate. Think of our classical, clockwork solar system. Given perfect knowledge of the current positions and movements of its planets and other bodies, we can predict their exact positions and movements at any time in the future with almost perfect accuracy. In the quantum world, however, uncertainty does away with any such ideas of perfect knowledge revealed by measurement. Its assertion that there are pairs of 'complementary' quantities such as position and momentum, where exact knowledge of one precludes knowing the other at all accurately, also undermines any concept of predictable cause and effect. If you cannot know the present in its entirety, you can have no idea what the future might bring.

Fuzzy logic

In the 1927 paper that introduced the uncertainty principle to the world, Werner Heisenberg established that there are pairs of quantities in the quantum world that cannot both be measured to an arbitrary level of precision at the same time.

One such pair is position and momentum – essentially a measure of a quantum particle's movement. If you know a particle's position x to within a certain accuracy Δ_x, then the uncertainty Δ_p on its momentum p is given by the mathematical inequality $\Delta_x \Delta_p \geq \hbar/2$. Here, $\hbar$ is a fixed number of nature, known as the reduced Planck constant. This inequality says that, taken together, Δ_x and Δ_p cannot undercut $\hbar/2$. Thus, in general, the more we know about where a particle is (the smaller Δ_x is), the less we can know about where it is (the larger Δ_p is), and vice versa.

The uncertainty principle also applies to other pairs of quantities, such as energy and time and the spins and polarizations of particles in various directions. The energy–time uncertainty relation is the reason why quantum particles can pop out of nothingness and disappear again. As long as the energy, Δ_E, that they borrow to do that and the time, Δ_t, for which they hang around don't break the uncertainty bound, the fuzzy logic of quantum mechanics remains satisfied.

The final proof of quantum weirdness

Since the 1930s, physicists have argued whether there was some deeper level of reality that could explain the strangeness of the quantum world. For example, entanglement is a serious blow to our conception of how the world works. In 1964 the Irish physicist John Bell showed just how serious. He worked out a mathematical way to tell whether a measurement on one quantum particle (a photon of light, say) truly could change the result of a measurement carried out on another particle immediately afterwards, or whether some invisible, non-quantum influence was responsible.

Bell's 'inequality' encapsulated the maximum correlation between the states of remote particles in experiments in which three 'reasonable' conditions hold: that experimenters have free will in setting things up as they want; that the particle properties being measured are real and pre-existing, not just popping up at the time of measurement; and that no influence travels faster than the speed of light, the cosmic speed limit.

As many experiments since have shown, quantum mechanics regularly violates Bell's inequality, yielding levels of correlation way above those possible if his conditions hold. The most recent, most watertight example of this sort of experiment was done in 2015 by a group led by Ronald Hanson at Delft University of Technology in the Netherlands.

It is worth taking a closer look at what they did and why. To understand, we have to go back to the 1930s, when physicists were struggling to come to terms with the strange predictions of the emerging science of quantum mechanics. The theory suggested that particles could become entangled, so that measuring one would instantly influence the measurement of the other, even if they were separated by a great distance. The implication was that particles could apparently communicate faster than any signal could pass between them. What's more, the theory also suggested that the properties of a particle are only fixed when measured, and prior to that they exist in a fuzzy cloud of probabilities.

Nonsense, said Einstein. He and others were guided by the principle of local realism, which broadly says that only nearby objects can influence each other and that the universe is 'real' – our observing it doesn't bring it into existence by crystallizing vague probabilities. They argued that quantum mechanics was incomplete, and that 'hidden variables' operating at some deeper layer of reality could explain the theory's apparent weirdness. On the other side, physicists like Niels Bohr insisted that we just had to accept the new quantum reality, since it explained problems that classical theories of light and energy could not.

Test it out

It was not until the 1960s that the debate shifted further to Bohr's side, thanks to the experimental tests that Bell's inequality allows.

A typical Bell test begins with a source of photons, which spits out two at the same time and sends them in different directions to two waiting detectors, operated by a hypothetical pair conventionally known as Alice and Bob. The pair have independently chosen the settings on their detectors so that only photons with certain properties can get through. If the photons are entangled according to quantum mechanics, they can influence each other and repeated tests will show a stronger pattern between Alice's and Bob's measurements than local realism would allow.

But what if Alice and Bob are passing unseen signals – perhaps through Einstein's deeper hidden layer of reality – that allow one detector to communicate with the other? Then you could not be sure that the particles are truly influencing each other in their instant, spooky quantum-mechanical way; instead, the detectors could be in cahoots, altering their measurements. This is known as the locality loophole, and it can be closed by moving the detectors far enough apart that there is not enough time for a signal to cross over before the measurement is complete. Physicists have done various tests to do just that, including shooting photons between two of the Canary Islands 143 kilometres apart.

Close one loophole, though, and another opens. The Bell test relies on building up a statistical picture through repeated experiments, so it doesn't work if your equipment doesn't pick up enough photons. Other experiments closed this detection loophole, but the problem gets worse the further you separate the detectors, as photons can get lost on the way. So moving the detectors apart to close the locality loophole begins to widen the detection one.

Hanson's team's test was the first experiment that simultaneously addressed both the detection loophole and the locality loophole.

Entangled diamonds

In their set-up, Alice and Bob sat in two laboratories 1.3 kilometres apart. Light takes 4.27 microseconds to travel this distance and the measurement took only 3.7 microseconds, so this was far enough to close the locality loophole.

Each laboratory had a diamond that contained an electron with a property called spin. The team hit the diamonds with randomly produced microwave pulses. This made them each emit a photon, which is entangled with the electron's spin. These photons were then sent to a third location, C, in between Alice and Bob, where another detector clocked their arrival time.

If photons arrived from Alice and Bob at exactly the same time, they would transfer their entanglement to the spins in each diamond. So the electrons would be entangled across the distance of the two labs – just what we need for a Bell test. What's more, the electrons' spin was constantly monitored, and the detectors were of high enough quality to close the detector loophole.

But the downside is that the two photons arriving at C rarely coincide – just a few per hour. The team took 245 measurements, so it was a long wait. The result was clear: the labs detected more highly correlated spins than local realism would allow. The weird world of quantum mechanics is our world (see Figure 2.6).

There is one remaining loophole for local realists to cling to, but no experiment can ever rule it out. What if there is some kind of link between the random microwave generators and

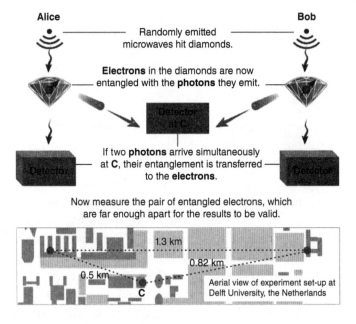

FIGURE 2.6 The first loophole-free experiment to prove quantum weirdness

the detectors? Then Alice and Bob might think they're free to choose the settings on their equipment, but hidden variables could interfere with their choice and thwart the Bell test.

Hanson's team note that this is a possibility, but assume that it is not the case. Other experiments have proposed using photons from distant quasars to produce random numbers, resulting in billions of years of separation.

None of this helps in the long run. Suppose the universe is somehow entirely predetermined, the flutter of every photon carved in stone since time immemorial. In that case, no one would ever have a choice about anything. As such, it's not really worth experimentalists worrying about – if the universe is predetermined, the complete lack of free will means that we have larger concerns.

What would Einstein have made of these results? Unfortunately, he died before Bell proposed his inequality, so we don't know whether subsequent developments would have changed his mind, but he would probably have been enamoured with the lengths people have gone to to prove him wrong.

Where does this loophole-free test leave us?

This loophole-free test of quantum weirdness pitches us into a philosophical dilemma. Do we not have free will, meaning that something somehow predetermines what measurements we take? That is not anyone's first choice. Are the properties of quantum particles not real – implying that nothing is real at all, but exists merely as a result of our perception? That is a more popular position, but it hardly leaves us any the wiser.

Or is there really an influence that travels faster than light? In 2008 the Swiss physicist Nicolas Gisin and his colleagues at the University of Geneva showed that, if reality and free will hold, the speed of transfer of quantum states between entangled photons held in two villages 18 kilometres apart was somewhere above 10 million times the speed of light.

Is there a size limit when objects stop behaving as waves?

According to the laws of quantum mechanics, wave–particle duality and quantum superpositions apply not only to the macroscopic world of electrons and atoms but to macroscopic objects too.

The boundary between the quantum world and the 'classical' everyday one has been weakening for years. In 1999 Anton Zeilinger and colleagues at the University of Vienna in Austria demonstrated that buckyballs – molecules of 60 carbon atoms – act like waves when they pass through gratings. And in 2003 the same team performed the feat with tetraphenylporphyrin, a large molecule related to chlorophyll, which set a new record for the heaviest object to show wave–particle duality.

Quantum effects have also nudged into the realm of objects visible to the naked eye. In 2010 researchers made a 0.06-millimetre-long supercooled metal strip simultaneously vibrate and not vibrate, putting it into a quantum superposition of states. The record is currently held by a cloud of 10,000 rubidium ions. Is there any limit to how large an object can be and still show quantum effects? Nothing in quantum mechanics says that this limit exists, but the more atoms an object has, the more likely those atoms are to interact with each other and their environment, destroying fragile quantum effects.

One of the key goals on the horizon is to superimpose objects on the scale of around 1 million atoms, says Vlatko Vedral, a quantum physicist at the University of Oxford. 'That is where something magical happens. That's the scale where life begins.' For him, a key experiment would be to fire an organism such as a virus at two slits under controlled conditions, as one of the interpretations of quantum mechanics is that living systems are the things that cause quantum superpositions to collapse. 'My bet is on the virus also being fully quantum mechanical,' he says. 'Give me enough research funding and I can probably interfere anything you like.'

3
What does it all mean?

Faced with the full-frontal assault of quantum weirdness, it is tempting to trot out the notorious quote from Nobel prizewinning physicist Richard Feynman (1918–88): 'Nobody understands quantum mechanics.' It does have a ring of truth to it, however: as classical beasts, we are ill equipped to see underlying quantum reality. Claiming to understand it can come with a heavy price – for instance, admitting to the existence of parallel universes.

An introduction to the multiverse

In 1911 the first world physics conference took place in Brussels, Belgium (see Chapter 1). The topic under discussion was how to deal with the strange new quantum theory and whether it would ever be possible to marry it to our everyday experience.

It is a question physicists are still wrestling with today. No experiment has ever disagreed with quantum theory's predictions, and we can be confident that it is a good way to describe how the universe works on the smallest scales. This leaves us with only one problem: what does it mean?

Claiming to understand quantum mechanics can come with a heavy price – for instance, admitting the existence of parallel universes. In this picture, the probabilistic wave functions that describe quantum objects do not 'collapse' to classical certainty every time you measure them; reality merely splits into as many parallel worlds as there are measurement possibilities. One of these carries you and the reality you live in away with it. In the words of Richard Feynman again, 'the "paradox" is only a conflict between reality and your feeling of what reality ought to be'.

Physicists try to answer this question with 'interpretations' – philosophical speculations, fully compliant with experiments, of what lies beneath quantum theory. No other theory in science has so many different ways of looking at it (see Figure 3.1). How so? And will any one win out over the others?

Take what is now known as the Copenhagen interpretation, for example, introduced by the Danish physicist Niels Bohr. It says that any attempt to talk about an electron's location within an atom, for instance, is meaningless without making a measurement of it. Only when we interact with an electron by trying to observe it with a non-quantum, or 'classical', device does it take on any attribute that we would call a physical property and therefore become part of reality. With its uncertainty principles and measurement paradoxes, the Copenhagen interpretation amounts to an admission that any attempt we make to engage with quantum reality reduces it to a shallow classical projection of its full quantum richness.

Then there is the many-worlds interpretation, where quantum strangeness is explained by everything having multiple existences in myriad parallel universes. Or you might prefer the de Broglie–Bohm interpretation, where quantum theory is considered incomplete: we are lacking some hidden properties that, if we knew them, would make sense of everything.

There are plenty more, such as the Ghirardi–Rimini–Weber interpretation, the transactional interpretation (which has particles travelling backwards in time), the British physicist Roger Penrose's gravity-induced collapse interpretation and the modal interpretation. In the last 100 years, the quantum zoo has become a crowded and noisy place and the debate about what quantum theory means shows no sign of going away. Nonetheless, only a few of these interpretations seem to matter to most physicists.

Wonderful Copenhagen

The most popular of all is Bohr's Copenhagen interpretation. Its popularity is largely due to the fact that physicists do not, by and large, want to trouble themselves with philosophy. Questions over what, exactly, constitutes a measurement, or why it might induce a change in the fabric of reality, can be ignored in favour of simply getting a useful answer from quantum theory.

That is why unquestioning use of the Copenhagen interpretation is sometimes known as the 'shut up and calculate' interpretation. This approach has a couple of disadvantages, however. It is never going to teach us anything about the fundamental nature of reality. That requires a willingness to look for places where quantum theory might fail rather than where it succeeds. Working in a self-imposed box also means that new applications of quantum theory are unlikely to emerge. The many perspectives we can take on quantum mechanics can be the catalyst for new ideas. Nowhere is this more evident than in the field of quantum information.

At the heart of this field is the phenomenon of entanglement, where the information about the properties of a set of quantum particles becomes shared between all of them. The result is that measuring a property of one particle will instantaneously affect the properties of its entangled partners, no matter how far apart they are.

Entanglement seemed such a strange idea that the physicist John Bell created a thought experiment to show whether it really could manifest itself in the real world (see Chapter 2). When it became possible to do, the experiment proved that it could, and it told physicists a great deal about the subtleties of quantum measurement. This result laid the foundations of quantum computing, where a single measurement could give you the answer to thousands, perhaps millions, of calculations done in parallel by entangled particles, and quantum cryptography, which protects information by exploiting the very nature of quantum measurement (see Chapter 4, 'Practical magic').

Both of these technologies have, understandably, attracted the attention of governments and industry keen to possess the best technologies – and to prevent them falling into the wrong

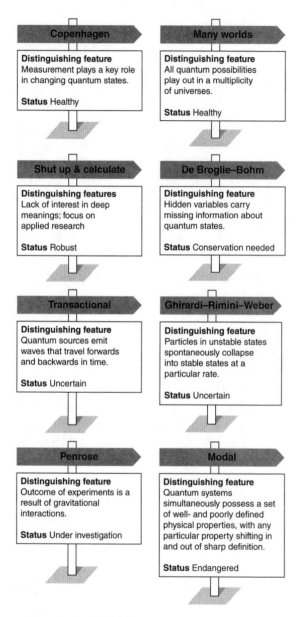

FIGURE 3.1 There is a 'zoo' of different interpretations of quantum theory

hands. Physicists, however, are more interested in what these phenomena tell us about the nature of reality. One implication of quantum information experiments seems to be that information held in quantum particles lies at the root of reality.

Adherents of the Copenhagen interpretation see quantum systems as carriers of information and measurement using classical apparatus as nothing special: it is just a way of registering change in the information content of the system. This new focus on information as a fundamental component of reality has also led some to suggest that the universe itself is a vast quantum computer.

However, for all the strides taken as a result of the Copenhagen interpretation, many physicists are critical of it. This is partly because it requires what seems like an artificial distinction between tiny quantum systems and the classical apparatus or observers that perform the measurement on them. Considering the nature of things on the scale of the universe has also provided Copenhagen's critics with ammunition. If the process of measurement by a classical observer is fundamental to creating the reality we observe, what performed the observations that brought the contents of the universe into existence?

Many worlds

The difficulty posed by this question is the reason why cosmologists now tend to be more sympathetic to an interpretation created in the late 1950s by Princeton University physicist Hugh Everett. His 'many-worlds' interpretation of quantum mechanics (see Figure 3.2) says that reality is not bound to a concept of measurement. Instead, the myriad different possibilities inherent in a quantum system each manifest in their own universe. David Deutsch, a physicist at the

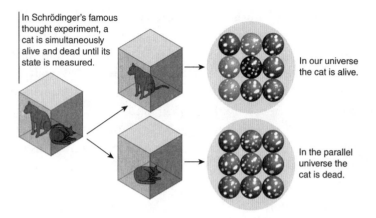

In Schrödinger's famous thought experiment, a cat is simultaneously alive and dead until its state is measured.

In our universe the cat is alive.

In the parallel universe the cat is dead.

FIGURE 3.2 The many-worlds interpretation of quantum mechanics, which suggests a continually branching series of multiverses

University of Oxford and the person who drew up the blueprint for the first quantum computer, says he can now think of the computer's operation only in terms of these multiple universes (see Interview, Chapter 5). To him, no other interpretation makes sense.

Not that the many-worlds interpretation is without its critics – far from it. Tim Maudlin, a philosopher of science based at Rutgers University in New Jersey, applauds its attempt to demote measurement from the status of a special process. At the same time, though, he is not convinced that many worlds provides a good framework for explaining why some quantum outcomes are more probable than others.

When quantum theory predicts that one outcome of a measurement is ten times more probable than another, repeated experiments have always borne that out. According to Maudlin, many worlds says that all possible outcomes will occur, given the multiplicity of worlds, but doesn't explain why observers still see the most probable outcome.

Unconventional Everett

Hugh Everett's many-worlds interpretation of quantum mechanics arose from what must have been the most world-changing drinking session of all time. One evening in 1954, in a student hall at Princeton University, graduate student Everett was drinking sherry with his friends when he came up with the idea that quantum effects cause the universe to constantly split. He developed the idea for his PhD thesis – and the theory held up.

But the leading physicists of Everett's day, in particular Niels Bohr, could not accept it. Everett had to publish a watered-down version of his idea. Thoroughly disgruntled, he left physics and joined the Pentagon, and worked in a team calculating potential deaths in the event of nuclear war. Everett's life was fascinating and tragic. He was a keen atheist, and before he died, aged 51, he left instructions for his widow Nancy to throw his ashes out with the garbage.

Deutsch believes that these issues have been resolved in recent times. However, his argument is abstruse and his claim has yet to convince everyone. Even more difficult to answer is what proponents of many worlds call the 'incredulous stare objection'. The obvious implication of many worlds is that there are multiple copies of you, for instance – and that Elvis is still performing in Vegas in another universe. Few people can bear this idea, but it may be just a question of getting used to this multiplicity of ourselves and others.

Deutsch thinks this will happen when technology starts to use the quantum world's stranger sides. Once we have quantum computers that perform tasks by being in many states at the same time, we will not be able to think of these worlds as anything other than physically real.

Hidden variables

Many worlds is not the only interpretation laying claim to cosmologists' attention. In 2008 Anthony Valentini of Imperial College London suggested that the cosmic microwave background radiation (CMB) that has filled space since just after the Big Bang might support the de Broglie–Bohm interpretation. In this scheme, quantum particles possess as yet undiscovered properties, dubbed hidden variables.

The idea behind this interpretation is that taking these hidden variables into account would explain the strange behaviours of the quantum world, which would leave an imprint on detailed maps of the CMB. Valentini says that hidden variables could provide a closer match with the observed CMB structure than standard quantum mechanics does. Though it is a nice idea, as yet there is no conclusive evidence that he might be on to something.

Other multiverses

It's not just quantum mechanics that leads to the inescapable conclusion that our universe is just a speck in a vast sea of universes. Different branches of physics spin off very different kinds of multiverses. Our current theory of how the universe came to be predicts an infinite expanse of other universes. If the Big Bang started with a period known as inflation – in which space itself expanded much faster than the speed of light so that there would be a multitude of other universes a lot like ours but causally disconnected from it – this could also potentially be the case with different arrangements of matter. Another level of multiverse arises from the theory of eternal inflation, when the space between universes continues to expand, and a limitless number of new 'bubble' universes with very different properties continue to form.

The other argument for a multiverse comes from string theory, which attempts to unify all known forces in physics. This maintains that all fundamental particles of matter and forces of nature arise from the vibration of tiny strings in 10 or 11 dimensions. For us not to notice the extra dimensions of space, they must be curled up, or compacted – so small as to be undetectable. For decades, mathematicians toiled over what different forms this compaction could take, and they found myriad ways of scrunching up space-time – between 10^{100} and 10^{500}. Each form gives rise to a different vacuum of space-time, and hence a different universe – with its own vacuum energy, fundamental particles and laws of physics.

As outlandish as these multiverses may seem, at least they are allowed to be real. Philosopher Nick Bostrom at the University of Oxford has upped the ante by arguing that the universe we experience is just a simulation running on an advanced civilization's supercomputer. The idea is that a long-lived civilization will develop essentially unlimited computing power and may choose to run multiple 'ancestor simulations' that will soon outnumber natural universes, making it plausible that we are one of them.

Testing the multiverse

Quantum Russian roulette is a thought experiment developed by physicist Max Tegmark to test the many-worlds hypothesis. It begins with an experimenter, a gun and photons (see Figure 3.3).

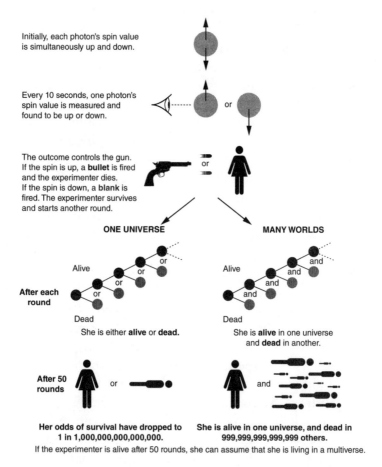

FIGURE 3.3 Max Tegmark's Russian roulette experiment

Interview: Parallel lives

Mark Everett (see Figure 3.4), known as E, and the creative force behind the rock band Eels, is the son of the physicist Hugh Everett, originator of the 'many worlds' view of quantum mechanics. E's band Eels has made a series of acclaimed albums, including Beautiful Freak, Electro-Shock Blues *and* Blinking Lights and Other Revelations. *His father, Hugh Everett, died in 1982, 25 years after proposing the 'many worlds' theory, which envisages multiple quantum states as constantly giving rise to parallel universes. Here E talks about the father he barely knew.*

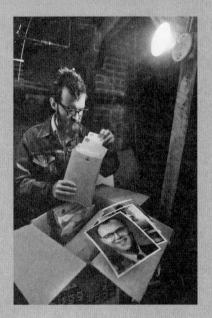

FIGURE 3.4 Mark Everett with a photo of his father, Hugh Everett

You lived with your father for 19 years, yet you say he was a stranger. Can you explain?

My father was always a physical presence, but he was just like a piece of furniture to me. When my sister was younger he might have been a bit more communicative, maybe. But I didn't see a lot of interaction with any of us. It was a weird and lonely childhood, because we were left to figure everything out on our own. You never get to feel like a child, and you learn everything the hard way. It was the sink-or-swim theory of child rearing.

Why do you think your father was so withdrawn?

He never faltered in his belief over his many-worlds theory, whereas no one else took it seriously. So he just gave up. That must be such an incredibly lonely feeling.

Yet by many measures your father was successful. He did classified research for the Pentagon and later became wealthy applying mathematical modelling in industry.

It's nice that he went on to find other success, but you can't help but wonder where he might have gone if he had gotten more encouragement in the world of physics.

Why didn't he get that support?

With relativity, Albert Einstein offered the world an appetizer before the main course, and that made it easier to swallow. My father just offered up the main course. It was hard for those on the Mount Rushmore of physics – Einstein and Niels Bohr – to say 'We're going to let this kid knock our faces off.' He went to Copenhagen in

1959, naively thinking that he was going to change Bohr's mind. To me, that's the defining moment of his life. After that he crawled into his shell.

Had he shown any signs of depression before that?

Well, the family has a strain of crazy in it: his mother had mental problems and my sister committed suicide. But the Copenhagen situation could take anyone who didn't have any problems and make them depressed.

Does knowing about his theory help you understand him as a man?

It makes it easier to let him off the hook for any short-comings he had as a father, because we're dealing with someone that's way above in terms of how his mind works. Einstein wasn't a great family man, either. These guys, I don't think they should be held to subscribe to normal rules.

His theory suggests that every choice we make spawns a series of parallel realities, which is a difficult idea. Has it affected your own outlook?

No, because I'm not a genius physicist. I've got my hands full with this world and I think in linear terms, as far as that goes. It can mean a lot of bad things are happening in the parallel universes – but if that's the way the world works, that's the way the world works.

Should I care about my other selves?

Every decision you make may spawn parallel universes where people are suffering because of your choice. Here, *New Scientist*'s Rowan Hooper explores the moral implications of the multiverse.

> 'The multiverse worries me. If the many-worlds interpretation is correct – and many physicists think it is – my actions shape the course not just of my life, but also of the lives of my duplicates in other worlds. Should I feel bad about the parallel Rowans that end up suffering as a result of my actions? How am I supposed to live with the knowledge that I am just one of umpteen Rowans in the multiverse, and that my decisions reach farther than I can ever know?
>
> 'You might think I should just ignore it. After all, the many-worlds interpretation says I'll never meet those other versions of me. So why worry about them? Well, most of us try to live by a moral code because we believe the things we do affect other people, even ones we'll never meet. We worry about how our shopping habits affect workers in distant countries; about as yet-unborn generations suffering for our carbon emissions. So why shouldn't we afford some consideration to our other selves?'

Max Tegmark, professor of physics at the Massachusetts Institute of Technology, understands Hooper's quandary. A leading advocate of the multiverse, he has thought long and hard about what it means to live in one. 'I feel a strong kinship with parallel Maxes, even though I never get to meet them. They share my values, my feelings, my memories – they're closer to me than brothers,' he says.

Cosmic perspective

Taking the cosmic perspective makes it difficult for Tegmark to feel sorry for himself: there is always another Max worse off than him. If he has a near miss while driving, he says he takes the experience more seriously than he did before he knew about the multiverse.

Nor can we avoid its consequences, says Hooper. Every time we make a decision that involves probability – such as whether to take an umbrella in case of rain – our decision causes the universe to branch. In one universe, we take the umbrella and stay dry; in another, we don't, and we get wet. The fundamental variability of the universe forces such choices upon us.

This is a momentous insight. We're living in a time akin to Copernicus realizing that Earth wasn't at the centre of the universe, or when Darwin realized that humans were not created separately from the other animals. Both of those realizations reshaped our conception of our place in the universe, our philosophy and our morality. The multiverse looks like the next great humbler of humanity.

Tough implications

The challenge for humans is to understand its implications – even physicists find this tough. For example, when Tegmark's wife was in labour with Philip, their eldest son, he found himself hoping that everything would go well. Then he admonished himself.

'It was going to go well, and it was going to end in tragedy, in different parallel universes. So what did it mean for me to hope that it was going to go well?' He couldn't even hope that the fraction of parallel universes where the birth went well was a large one, because that fraction could in principle be calculated. 'So it doesn't make any sense to say, "I'm hoping something about this number." It is what it is.'

Hope, it turns out, is the next casualty of the multiverse. You make a decision, and you end up on a branch of the multiverse with a 'good' outcome, or you find yourself on a 'bad' branch. You cannot wish your way on to a good one. This is tough: how are we meant to live without hope?

Perhaps a philosopher can help us take a broader perspective. David Papineau, of King's College London, offers this:

'Say you put your money on a horse which you think is a very good bet. It turns out that it doesn't win, and you lose all your money. You think, "I wish I hadn't done that." But you brought benefits to your cousins in other universes where the horse won. You've just drawn the short straw in finding yourself in the universe where it lost. You didn't do anything wrong. There's no sense that the action you took earlier was a mistake.'

Of course, 'I didn't make a mistake' would not impress our loved ones if we had bet all our savings on a horse and found ourselves on the 'wrong' branch. But then, that wouldn't be the sensible thing to do – and one of the great attractions of Everett's interpretation, according to Papineau, is that it's not 'messy' as long as you act rationally.

With orthodox thinking, there are two ways of evaluating risky actions, he explains. First, did you make the choice that was most in line with the odds? If we needed money, and my stake had been proportionate, it might have been. Second, did it work out well? There are any number of reasons it might not – the horse might fall, or just defy the odds and trail in last.

It offends Papineau that these two ways of being 'right' – choosing wisely and getting lucky – do not go hand in hand. 'The idea that the right thing to do might turn out to have been the wrong thing seems to me to be a very ugly feature of orthodox thinking,' he says. This doesn't arise in the many-worlds interpretation, where every choice is made and every outcome occurs. That leaves no place for hope or luck, but nor does it leave room for remorse. It's an elegant, if cold-blooded, way to look at things.

This elegance has always been part of the multiverse's appeal (see Figure 3.5). In quantum mechanics, every object in the universe is described by a mathematical entity called a wave function, which describes how the properties of subatomic particles can take several values simultaneously. The trouble is

FIGURE 3.5 An artistic interpretation of the multiverse: members of Spain's La Fura dels Baus Theatre and Ukraine's Center of Contemporary Art 'DAKH' perform 'Multiverse' at an art festival in Kiev, 4 September 2010.

that this fuzziness vanishes as soon as we measure any of those properties. The original explanation for this – the so-called Copenhagen interpretation – says the wave function collapses to a single value whenever a measurement is made.

Multiple outcomes

Hugh Everett called this enforced separation of the quantum world from the everyday, classical one a 'monstrosity', and decided to find out what happened if the wave function did not collapse. The resulting mathematics showed that the universe would split every time a measurement is made – or, in human terms, whenever we make a decision with multiple possible outcomes.

For Don Page, a theoretical physicist at the University of Alberta in Edmonton, Canada, this elegance goes far beyond human actions. Page is both a hard-core Everettian and a committed Christian. Like many modern physicists, he agrees with Everett's stance that collapsing the wave function is unnecessarily complicated. What's more, for Page it has a happy side effect: it explains why his god tolerates the existence of evil.

'God has values,' he says. 'He wants us to enjoy life, but he also wants to create an elegant universe.' To God the importance of elegance comes before that of suffering, which, Page infers, is why bad things happen. 'God won't collapse the wave function to cure people of cancer, or prevent earthquakes or whatever, because that would make the universe much more inelegant.'

For Page, that is an intellectually satisfying solution to the problem of evil. And what's more, many worlds may even take care of free will. Page doesn't actually believe we have free will, because he feels we live in a reality in which God determines everything, so it is impossible for humans to act independently. But in the many-worlds interpretation every possible action is actually taken. 'It doesn't mean that it's fixed that I do one particular course of action. In the multiverse, I'm doing all of them,' says Page.

There are limits to Page's willingness to leave his fate to the multiverse, however. He was once offered $1 million to play quantum Russian roulette, which is a good game for a multiverse aficionado: you cannot lose (see 'Testing the multiverse' above). Page thought about it, then declined: he disliked the thought of his wife's distress in the worlds where he died.

It is something of a relief to find that even many-worlds experts ultimately behave in much the same way as people who know nothing of it. But their knowledge shapes the way they think about their decisions. Perhaps it is more natural for us to think about how our actions affect our 'other selves' than about the arid probabilities of risk and reward.

If anyone is going to buck this trend, it must be David Deutsch, probably the most hardcore of Everettians. Surely he can give the last word on what it means to live in the multiverse. He does, but the answer is unexpected:

'Decision theory in the multiverse tells us that we should value things that happen in more universes more, and things that happen in fewer universes less. And it tells us that the amount by which we should value them more or less is, barring exotic circumstances, exactly such that we should behave as if we were valuing the risks according to probabilities in a classical universe.'

So the right thing to do remains the right thing to do.

Of course, Deutsch's approach could be wrong, a possibility he accepts, although he is adamant that the multiverse exists. But if he is right, his conclusion only reinforces what his peers have been saying: the best way to live in the multiverse is to think carefully about how you live your life in this one.

Does consciousness create reality?

With its multiverses and cats both alive and dead, quantum mechanics is certainly weird. But some physicists have proposed that reality is even stranger that we might think: the universe only becomes real when we look at it.

This version of the anthropic principle – known as the participatory universe – was first put forward by John Archibald Wheeler, a leading light of twentieth-century physics. He likened what we call reality to an elaborate papier-mâché construction supported by a few iron posts. When we make a quantum measurement, we hammer one of those posts into the ground. Everything else is imagination and theory.

For Wheeler, however, making a quantum measurement not only gives us an objective fix on things but also changes the course of the universe by forcing a single outcome from many possible ones. In the famous double-slit experiment, for example, light is observed to behave either as a particle or as a wave, depending on the set-up (see Chapter 2). The most baffling thing is that photons seem to 'know' how and when to switch. But this assumes that a photon has a physical form before we observe it. Wheeler asked: what if it doesn't? What if it takes one only at the moment we look?

Even the past may not yet be fixed. Wheeler proposed a cosmic version of the double-slit experiment, in which light from a quasar a billion light years away reaches us by passing around a galaxy that distorts its path, producing two images, one on either side of the galaxy. By pointing a telescope at each, observers would see photons travelling one of two routes as particles. But by arranging mirrors so that photons from both routes hit a detector at the same time, they would see light arrive as a wave. This time, the act of observation reaches across time to change the nature of the light leaving the quasar a billion years ago.

For Wheeler, this meant that the universe could not really exist in any physical sense – even in the past – until we measured it. And what we do in the present affects what happened in the past – in principle, all the way back to the origins of the universe. If he is right, then to all intents and purposes the universe didn't exist until we and other conscious entities started observing it.

So peculiar is the quantum world that the idea that it only exists because of us seems almost sensible.

How can we understand quantum reality?

'I believe in an external physical reality beyond my own experience,' says Johannes Kofler of the Max Planck Institute of Quantum Optics in Garching, Germany. 'The world would be there without me, and was there before me, and will be there after me.'

Given what we know about quantum physics, that seems a bold statement. The assaults that this most fundamental theory of reality makes on our intuition are legion. For Aephraim Steinberg at the University of Toronto in Canada, dealing with such troubling concepts is a matter of retraining our brains. 'As much as we talk about the "counter-intuitiveness" of quantum mechanics, we just mean that it is counter to the intuitions we have before we learn quantum mechanics,' he says. After all, we are not very good at second-guessing aspects of classical reality, either: how many of us would naturally say that feathers and bricks fall at the same rate under gravity?

With quantum physics, though, it does not help that the quantities used to describe objects seem to exist only mathematically. Visualizing a wave function as a real thing is fine for a single particle, but things rapidly get more complicated. 'Once you're talking about more than one particle, the wave function lives in some high-dimensional space I don't know how to visualize,' says Steinberg. He has to break a complex quantum system down into parts. 'But they're all merely ways of chipping away at the abstract mathematical object I know provides the complete description.'

More fundamentally, though, if we accept quantum physics at face value then at least one of two dearly held principles from the classical world must give. One is realism, the idea that every object has properties that exist without our measuring them. The other is locality, the principle that nothing in the universe can influence anything else 'instantaneously' – faster than the speed of light.

For most quantum physicists, it is realism that has to give, given all the evidence that the cosmic speed limit is never broken.

4
Practical magic

With all its strange properties, quantum mechanics might seem to have little relevance to the everyday world, but quite the opposite is the case. Quantum theory has transformed the world through its spin-offs. Almost all modern gadgets – computers, mobile phones, games consoles, cars – contain memory chips based on the transistor, whose operation relies on the quantum mechanics of semiconductors. Lasers, which have widespread applications in data storage, printing, communication and manufacturing, are also based on quantum properties.

Transistors: the hole story

You are surrounded by transistors. They are in computers, in your phone and in your household devices. Around 3 sextillion (that's 3×10^{21}) have been made since the technology was first created in 1947 – that's 428 billion for every person on Earth.

A transistor is essentially a lump of semiconducting material sitting between two electrodes that acts as a switch. A further electrode supplies a pulse of voltage, 'opening' the switch and allowing current to flow through the transistor. Transistors are used not only to amplify electrical signals, such as the radio signals picked up by an aerial, but also as electronic switches. Networks of these switches can form logic circuits, which control electronic appliances or manipulate information in your computer.

Ultimately, transistors depend on manipulating the way electrons jump between different energy levels in the atoms in semiconducting materials. This process, at a fundamental level, is based on quantum behaviour.

Today, new types of quantum transistors are in development. In 2015, for example, researchers demonstrated that two silicon transistors acting as quantum bits can perform a tiny calculation.

The discovery of transistors

The transistors that hum away in computer processors today depend on the qualities of that odd half-breed of material known as a semiconductor. Sitting on the cusp of electrical conduction and insulation, semiconductors sometimes let currents pass and sometimes resolutely block their passage.

By the early twentieth century some aspects of this dual personality were well documented. For example, the semiconductor galena, or lead sulphide, was known under certain circumstances to form a junction with a metal through which current travelled in only one direction. This had made it briefly popular in early wireless receivers to transform oscillating radio signals into steady direct current. However, this was a time-consuming and sometimes infuriating business, symptomatic of all semiconductors' failings. There seemed no logical explanation for their properties; a slight change in temperature or their material make-up could tip them from conduction to insulation and back again. It was tempting to think that their caprices might be tamed to make reliable, reproducible electrical switches, but no one could see how.

And so in the radio receivers and telephone and telegraph systems of the 1920s and 1930s vacuum tubes came to reign supreme – despite being bulky, failure-prone and power-hungry. However, the seeds of their demise and semiconductors' eventual triumph were being sown.

In 1928 Rudolf Peierls, a young Berlin-born Jew, was working as a student of the great pioneer of quantum physics, Werner Heisenberg, in Leipzig, Germany. The convolutions of history would later make Peierls one of the UK's most respected physicists, and pit him against his mentor in the race to develop the first atomic bomb. At the time, though, he was absorbed by a more niggling problem: why were electrical currents in some metals deflected the wrong way when they hit a magnetic field?

Absences of electrons

To Peierls the answer was obvious. 'The point [was] you couldn't understand solids without using the quantum theory,' he recalled in a 1977 interview. Just as quantum theory dictates that electrons orbiting an atom could not have just any energy, but are confined to a series of separate energy states, Peierls showed that, within a solid crystal, electrons are shoe-horned into 'bands' of allowed energy states. If one of these bands had only a few occupied states, electrons had great freedom to move, and the result was a familiar electron current. But if a band had only a few vacant states, electron movement would be restricted to the occasional hop into a neighbouring empty slot. With most electrons at a standstill, these vacancies would themselves seem to be on the move: mobile 'absences of electron' acting for all the world like positive charges – and moving the wrong way in a magnetic field.

Although Peierls's band calculations were the germ of a consistent, quantum-mechanical way of looking at how electrical conduction happened, no one quite made the link at the time. That didn't happen until 1940, when a team at Bell Labs led by engineer Russell Ohl attempted to tame the properties of the semiconductor silicon. At the time, silicon's intermittent conduction was thought to be the result of impurities in its crystal structure, so Ohl and his team set about purifying it. One day, a glitch in the purification process produced a silicon rod with a truly bizarre conducting character. One half acted as if dominated by negatively charged carriers, electrons. The other half, though, seemed to contain moving positive charges.

That was odd, but not half as odd as what happened when Ohl and his team lit up the rod. Left to its own devices, the imbalanced silicon did nothing at all. When a bright light was shone on it, however, it flipped into a conducting state, with current flowing from the negative to the positive region. A little more probing revealed what was going on. Usually, a silicon atom's four outer electrons are all tied up in bonds to other atoms in the crystal. But on one side of Ohl's rod, a tiny impurity of phosphorus with its five outer electrons was creating an

In pure silicon, all electrons that might conduct are tied up as bonds between atoms. That changes with the addition of an impurity such as boron.

Boron atoms have one fewer electron than silicon atoms, which creates 'holes' in the electron structure. Neighbouring electrons can leap into these holes, leaving a fresh hole behind them.

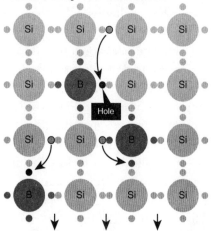

Direction of travel of electrons, or negative charge

While most electrons remain bound, the net effect is a movement of holes in the opposite direction to any electron current.

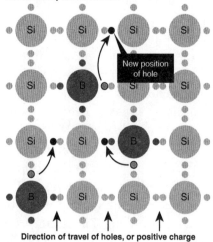

Direction of travel of holes, or positive charge

FIGURE 4.1 The transistors that run our computers are powered by holes.

excess of unattached electrons. On the other, a small amount of boron with just three electrons was causing an electron deficit (see Figure 4.1).

When kicked into action by the light, electrons were spilling over from the region of their excess to fill the holes in the electron structure introduced by the boron. Ohl named his discovery the positive-negative or 'p-n' junction, owing to its two distinct areas of positive and negative charge carriers. Its property of converting light energy into electric current made it, incidentally, the world's first photovoltaic cell.

Swarming holes

A few years later William Shockley, a physicist at Bell Laboratories in Murray Hill, New Jersey, heard about Ohl's breakthrough. It did not take him long to spot the p-n junction's potential. He was fascinated by the thought that, by pressing a metal contact to the junction's midpoint, you might use an external electric field instead of light to control the current across it. In a sufficiently thin layer of n or p-type silicon, he reasoned, the right sort of voltage would make electrons or holes swarm towards the contact, providing extra carriers of charge that would boost the current flow along the surface layer. The result would be an easily controllable, low-power, small-scale amplifier that would make the vacuum tube obsolete.

His first attempts to realize the dream, though, were unsuccessful. 'Nothing measurable, no measurable results,' he noted of an early failure. But at this point Shockley was obliged to leave further investigations to two of his highly qualified subordinates, John Bardeen and Walter Brattain. It proved a frustrating chase, and it was a classic combination of experimental nous and luck that led the pair to success – plus Bardeen's spur-of-the-moment decision to abandon silicon for its slightly more predictable semiconducting sister germanium. This finally produced the right sort of amplification effect, boosting the power of input signals, sometimes by a factor of hundreds. Just one thing didn't add up: the current was moving through the device in the wrong direction.

No doubt they would have solved this conundrum, given time, but Shockley stole a march on his colleagues. There was a way out of the impasse, he realized, and the answer lay in the holes. What if they were real, like a particle, not just the absence of an electron, and could happily coexist with electrons? That would explain what was going on in the transistor.

Shockley used the idea to develop a transistor that exploited the independence of electrons and holes. This was the 'p-n-p' transistor, in which a region of electron excess was sandwiched between two hole-dominated areas. Apply the right voltage and the resistance of the middle section could be broken down, allowing holes to pass through hostile electron-populated territory without being swallowed up. It also worked in reverse: electrons could be made to flow through a central region given over to holes. This was the principle that came to underpin the workings of commercial transistors in the decades that followed.

1.3×10^{12}
Transistors in use globally in 1970

The rest, as they say, is history. For Shockley, it was not a happy one. He did not at first tell Bardeen and Brattain of his new course, and even attempted to claim sole patent rights over the first transistor. The relationship between the three men never recovered. By the time they shared the Nobel Prize in Physics for their discovery in 1956, Shockley had left Bell Labs to form the Shockley Semiconductor Laboratory to capitalize on his transistor alone. But his high-handed and increasingly paranoid behaviour soon led to a mass mutiny from the bright young talents he had hired, such as Gordon Moore and Robert Noyce, who went on to found Intel, which remains the world's largest manufacturer of microchips.

1.4×10^{20}
Transistors in use globally in 2010

What happens to quantum mechanical properties at absolute zero?

The weird effects of the quantum world extend their reach at low temperatures. In everyday solids, liquids and gases, heat or thermal energy arises from the motion of atoms and molecules as they zing around and bounce off each other. But at very low temperatures, the odd rules of quantum mechanics reign. Molecules do not collide in the conventional sense; instead, their quantum mechanical waves stretch and overlap. When they overlap like this, they sometimes form a so-called Bose–Einstein condensate, in which all the atoms act identically like a single 'super-atom'. The first pure Bose–Einstein condensate was created in Colorado in 1995 using a cloud of rubidium atoms cooled to less than 170 nanokelvin.

But absolute zero (0 kelvin or -273.15 °C) is an impossible goal to reach. Practically, the work needed to remove heat from a gas increases the colder you get, and an infinite amount of work would be needed to cool something to absolute zero. In quantum terms, you can blame Heisenberg's uncertainty principle, which says that the more precisely we know a particle's speed, the less we know about its position, and vice versa. If you know your atoms are inside your experiment, there must be some uncertainty in their momentum keeping them above absolute zero – unless your experiment is the size of the whole universe.

Lasers

Lasers might be more than half a century old, but they are still the youthful pin-ups of fundamental physics. Since the first one was unveiled in 1960, their application can be seen everywhere, from cutting and welding to combating cancer and cataracts, to powering telecoms and consumer electronics. Advances in the laser lab translate into gadgets in our homes at astonishing speed: think of the progression from CD to DVD and now Blu-ray technology, in just a few decades.

Einstein took the first step to the eventual development of the laser by suggesting in 1917 that atoms could be stimulated to emit light. He was proved right a decade later, but it was not until 1954 that Charles Townes and other scientists at Columbia University in New York built the first 'maser' that produced a microwave beam. The development of the maser spurred Townes and others to try to extend the idea to visible and infrared light.

By 1957 Townes and Arthur Schawlow, then at Bell Telephone Laboratories, analysed how to make an 'optical maser'. Meanwhile, Gordon Gould, then a 37-year-old graduate student at Columbia, filled his notebooks with similar ideas from what he called a 'laser'. Gould sought a patent. People still argue about who had the laser idea first. Townes's work earned him a share in the 1964 Nobel Prize in Physics.

The Townes–Schawlow paper stimulated many efforts to make lasers. However, the winner of the great laser race was a little-known young American physicist, Theodore Maiman. On 15 May 1960 he slipped a small ruby rod, with silvered ends, into a spring-shaped flashlight. When he fired the flashlight, the ruby rod emitted a bright pulse of deep-red light – the first laser beam.

The field has moved on quickly since then. Today, by carefully selecting the elements in laser diodes, a range of different wavelengths can be emitted, leading to a range of different applications (see Figure 4.2).

The next stage in laser evolution is also under way, a sea change in how laser light is produced. A new wave of devices aims to exploit particle-like packets of energy to produce their light – packets that are neither light nor matter, but both.

Not so weird

Lasers and transistors exploit quantum particles such as electrons and protons, but they do not directly harness the weird quantum behaviours of superposition and entanglement. This is the promise of the field of quantum information (see Chapter 5).

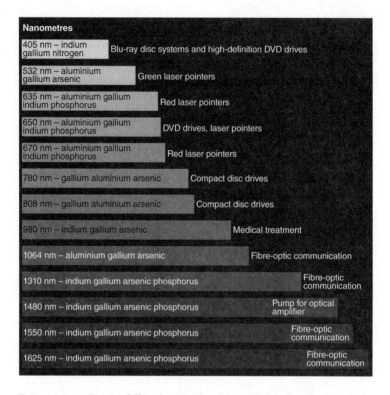

Nanometres	
405 nm – indium gallium nitrogen	Blu-ray disc systems and high-definition DVD drives
532 nm – aluminium gallium arsenic	Green laser pointers
635 nm – aluminium gallium indium phosphorus	Red laser pointers
650 nm – aluminium gallium indium phosphorus	DVD drives, laser pointers
670 nm – aluminium gallium indium phosphorus	Red laser pointers
780 nm – gallium aluminium arsenic	Compact disc drives
808 nm – gallium aluminium arsenic	Compact disc drives
980 nm – indium gallium arsenic	Medical treatment
1064 nm – aluminium gallium arsenic	Fibre-optic communication
1310 nm – indium gallium arsenic phosphorus	Fibre-optic communication
1480 nm – indium gallium arsenic phosphorus	Pump for optical amplifier
1550 nm – indium gallium arsenic phosphorus	Fibre-optic communication
1625 nm – indium gallium arsenic phosphorus	Fibre-optic communication

FIGURE 4.2 By carefully selecting the chemical elements used to make
a laser diode, as well as their relative proportions, engineers can construct
semiconductor lasers that emit at wavelengths from the blue to the infrared
part of the spectrum, making them suitable for a variety of applications.

5
Quantum information and computing

Processing information in quantum states, rather than in the electrical currents of conventional computer chips, offers the prospect of peerlessly powerful, economical and secure number crunching. That is the theory, at least. The challenge is to make it a reality. Looking to the future, quantum cryptography promises a new way to communicate in absolute privacy – vital for an increasingly networked society. New uses for quantum mechanics are arriving all the time.

What makes quantum computers so different?

The idea of harnessing the laws of quantum mechanics to build a computer was first proposed by the physicist Richard Feynman (see Figure 5.1) in 1982 and the physicist David Deutsch produced the first theoretical blueprint in 1985. The field flourished in the decades that followed (see the timeline later in this chapter). But how do you actually build a quantum computer, and how do they work? Here is a rundown of the basics.

Conventional, classical computers process information using the presence or absence of electrical charge or current. These classical bits have two positions, on (1) and off (0). Semiconductor switches – transistors – flip these bits, making logic gates such as AND, OR and NOT. Combining these gates, we can compute anything that is in principle computable.

In quantum computation, the switching is between quantum states. Quantum objects can generally be in many states at once: an atom may simultaneously be in different locations or energy states, a photon in more than one state of polarization, and so on. In essence, a quantum bit, or qubit, is in a 'superposition' that stores 0 and 1 at the same time.

This already suggests an enhanced computational capacity, but the real basis of a quantum computer's power is that the states of many qubits can be entangled in one another. This creates a superposition of all the possible combinations of the single-qubit states. Different operations performed on different parts of the superposition at the same time effectively make a massively powerful parallel processor. The power increase is exponential: n qubits have the information-processing capacity of 2^n classical bits (see Figure 5.2). A 400-qubit quantum computer would be like a classical computer with 10^{120} bits – far more than the number of particles estimated to exist in the universe.

FIGURE 5.1 Richard Feynman, pioneer in the field of quantum computing

The qubit

Ordinary computers use 'bits' to process information. The basic unit of quantum computing is the qubit. These are physical systems that can exist in two different states, so they can represent the 1s and 0s that make up the binary code that computers run. A qubit might be an electron held in a magnetic field or a photon that is polarized so its spin can be easily manipulated. Preparing qubits, as well as reading and writing to them, involves some dedicated hardware. This can be based on a ruby laser, a non-linear crystal or even a pink diamond.

Superposition

The extraordinary advantage a qubit has over a normal bit is that it can be put into a superposition state, being 0 and 1 at the same time. But this is tricky to pull off: any stray heat, electromagnetic noise or physical bump can knock it out again. To prevent this entails serious refrigeration or a state-of-the-art vibration containment system. Even then, you can only run the computer for a limited time before the superposition collapses. This 'coherence time' is very important.

Superposition

One qubit encodes **0 and 1** simultaneously.

Entanglement

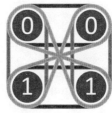

Two qubits store all four permutations simultaneously **00, 01, 10, 11**.

Similarly, three qubits store eight states, four qubits 16 states and so on – which gives an exponential increase in computing power.

Teleportation

Alice and Bob share a pair of entangled qubits.

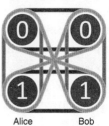

Alice   Bob

To transmit a 2-bit classical message **(00, 01, 10 or 11),** Alice measures her qubit, collapsing the superposition into one of its four possible states.

Alice   Bob

Bob sees the consequence of the measurement in his own qubit, so knows which bits Alice had – without anything physical being transferred.

FIGURE 5.2 Quantum superposition and entanglement combine to allow information to be processed more efficiently and teleported over distances.

Entanglement

This is where the 'magic' really happens. When two subatomic particles become inextricably interlinked, or entangled, this link lets you manipulate multiple qubits at once. That's what makes quantum computers so remarkable: just eight qubits, held in superposition and entangled, can simultaneously represent every number from 0 to 255, letting you carry out many operations at once. For quantum computation, an important quality is how many entangled qubits the machine can manage at once. At the moment, 14 is the record, achieved in 2011 by Rainer Blatt's group at the University of Innsbruck in Austria.

Error correction

Even normal computers make mistakes. Sometimes a voltage spike or a passing cosmic ray can buffet a bit, changing it from a 0 to a 1, say. Processors deal with this by keeping copies, but this is not an option for qubits, thanks to a law called the no-cloning theorem.

Fortunately, there are error-correction algorithms to get around this. The drawback is that these need a large number of qubits, anything between 100 and 10,000 times as many as are needed for the actual computation you are trying to perform. Happily, our ability to assemble arrays of qubits for error correction has come on in leaps and bounds. And error rates have been creeping downwards, too. In June 2014 IBM unveiled error-correcting code that is well suited to the large arrays of qubits expected to outperform regular machines.

What makes for a good qubit?

In 1997 David DiVincenzo of IBM wrote down some desirable conditions that remain a rough, though not exhaustive, check-list for what any practical quantum computer must achieve:

- **Scalability**
To outgun a classical computer, a quantum computer must entangle and manipulate hundreds of qubits. Quantum computers built so far have just a handful. Scaling up is a big hurdle: the larger the system, the more prone it is to 'decohere' in the face of environmental noise, losing its essential quantumness.

- **Initialization**
We must be able to reliably set all the qubits to the same state (to zero, say) at the beginning of a computation.

- **Coherence**
The time before decoherence kicks in must be a lot longer than the time to switch a quantum logic gate – preferably, several tens of times. In most practical implementations so far, this requires an operating temperature near absolute zero, to limit the effects of environmental interference.

- **Accuracy**
The results of manipulations must be reproduced accurately by the qubit, even when many manipulations are applied in sequence.

- **Stable memory**
There must be a reliable way to set a qubit's state, keep it in that state, and reset it later.

Number crunching

As well as all this quantum trickery, the promise of quantum computers rests largely on two algorithms. One, developed in 1994 by Peter Shor, then of Bell Laboratories, provides a way for a quantum computer to speedily find the prime factors of large numbers (see the interview later in this chapter). Classical computers effectively have to try to divide the given number by all possible prime factors (2, 3, 5, 7, 11 and so on) in turn, whereas quantum computers can do these divisions simultaneously. Conventional encryption methods rely on the fact that classical computers cannot factorize efficiently.

If Shor's algorithm were ever implemented on a large scale, encrypted information such as the PIN for your bank card would be vulnerable to hacking – and quantum cryptography would be the only viable defence (see later in this chapter). There is no need to worry just yet, as quantum computers are not sufficiently powerful. In the longer term, an algorithm devised by physicist Lov Grover in 1996, also at Bell Labs, may become a quantum computer's greatest selling point. This provides a recipe by which a quantum computer can radically speed up how we access and search large bodies of data.

Take the example of a database listing the contents of a library. Searching this database for a particular book with a classical computer takes a time that scales with the number of books, n; Grover's algorithm shows that for a quantum computer it scales with $\sqrt{n}$. For a library of a million books, this amounts to 1,000 times faster. Implementing such an algorithm has ubiquitous appeal: almost all computationally hard problems – for instance that of the travelling salesman who has to find the shortest route between a number of different cities – ultimately reduce down to a search for the optimal solution. There is, however, a way to go before this can be achieved.

Hardware

There are many ways of making the 'qubits' for a quantum computer, from polarizing light or cooling atoms to taming the collective motion of electrons. But qubits must fulfil some stringent criteria, particularly in proving robust, or 'coherent', in the face of buffeting from their surrounding classical environment. No single sort of qubit has yet ticked all the boxes. Superconducting qubits have been around longer, but spin is the ultra-cold new thing – and there are in addition a few wild-card options, as explained below.

Superconducting qubits

The forerunner of all quantum computer technology occurred in 1962, when Cambridge physicist Brian Josephson showed that putting a small gap into a strip of superconductor – a material with zero resistance to the flow of electricity at low temperatures – has a surprising effect. For example, superconducting loops incorporating such a 'Josephson junction' let current flow clockwise and anticlockwise simultaneously. That's a superposition of states – just what you need for a qubit. What's more, these systems are manufactured on the mainstay material of the technology industry: silicon.

Superconducting quantum interference devices (SQUIDs) already exploit this effect to make incredibly sensitive measurements of electromagnetic fields. But electron movements and magnetic field states within a SQUID can also be manipulated using external fields to form the bits of a quantum logic device. SQUID qubits offer good initialization and decoherence times, typically ten or so times greater than the time taken to switch a gate. With large numbers of qubits, however, heating due to the external fields used for manipulation becomes an issue.

The company D-Wave Systems in Burnaby, British Columbia, Canada, claimed in 2011 to have developed a 128-qubit computer, and bigger ones since then. The latest, announced in January 2017 (see also 'Want one today?' below), is the D-Wave 2000Q, with 2,000 qubits; but it remains controversial whether this device is fully quantum. IBM already has a 5-qubit superconducting computer that can execute quantum computation, and it is hooked up to the Internet for others to use. Google has a 9-qubit computer and in the near future hopes to have a 49-qubit one (using a grid of 7 × 7 qubits).

Spin qubits

Nuclear spin states manipulated using magnetic fields were among the first qubits explored. The great advantage of spin states is that they make qubits at room temperature, albeit with a very low initialization accuracy of about one in a million. But the disrupting effects of thermal noise on entanglement mean that nuclear-spin computers are limited to about 20 qubits before their signal becomes washed out. A variant on the spin theme exploits nitrogen impurities in an otherwise perfect diamond (carbon) lattice. These introduce electrons whose spin can be manipulated electrically, magnetically or with light – but scaling up to anything more than a couple of spins has proved difficult.

Ion-trap quantum computing

Trapping ions is perhaps the most advanced method of making a quantum computer's qubits. Positively charged ions are caught in electromagnetic fields and cooled to a nanokelvin or so, to reduce their vibrations and limit decoherence. Information is then encoded in the ions' energy levels and manipulated using laser light. That brings excellent initialization success (99.99 per cent), accuracy (over 99 per cent) and stable memory storage (years).

In 1995 David Wineland and his colleagues at the US National Institute of Standards and Technology in Boulder, Colorado, used trapped ions to create the first quantum logic gate – a controlled NOT (C-NOT) gate for disentangling entangled ions. In 2011 physicists from the University of Innsbruck, Austria, developed a 6-qubit trapped-ion computer that fulfilled the specifications for a universal quantum simulator that Richard Feynman had set out in 1981. And in 2016 physicists from the National Institute of Standards and Technology in Boulder, Colorado, trapped a record-breaking 219 beryllium ions and entangled their quantum properties with lasers. Decoherence and scalability remain interrelated problems, however.

Photonic quantum computing

Photons look as though they would make good qubits: they are easily superposed and stay coherent for good lengths of time. The position, polarization or just number of photons in a given space can be used to encode a qubit. Though initializing their states is easy, photons are slippery: they are easily lost and do not interact very much with one another. That makes them good for communicating quantum information, but to store that information we need to imprint photon states on something longer-lived, such as an atomic energy state.

If we can do that, quantum computing with photons is a promising concept, not least because the processing can be done at room temperature. In 2012 a team at the University of Vienna, Austria, used four entangled photons to perform the first blind quantum computation. Here a user sends quantum-encoded information to a remote computer that does not itself 'see' what it is crunching. This may be a future paradigm – totally secure quantum cloud computing. Records continue to be broken for the greatest number of entangled protons. In 2016, for example, a team at the University of Science and Technology of China developed a laser system that produced ten entangled photons.

Cold atoms

Collections of many hundreds of atoms might make for good qubits when trapped, cooled and arranged using lasers in a two-dimensional array known as an optical lattice. The energy states of these atoms can encode information that can be manipulated using further lasers, as with trapped ions. We have mastered the basic techniques, but making a true quantum computer from cold atoms awaits establishing reliable entanglement among these aloof bodies.

Atom–light hybrids

Cavity electrodynamics is a quantum computing approach that aims to combine stable cold atoms with agile photons. Light is trapped inside a micrometre-scale cavity and atoms sent flying through, with logical operations performed through the interactions of the two.

Initialization is highly efficient, and the decoherence time allows ten or so gate operations to be performed – although scaling the technology up awaits reliable ways of entangling trapped cold atoms. Serge Haroche of the Collège de France in Paris, one of the pioneers of this approach, shared the 2012 Nobel Prize in Physics with trapped-ion researcher David Wineland.

Topological quantum computing

In this technique, qubits are encoded in the way subatomic particles move past one another. However, this promising basis for a quantum computer has yet to get off the theoretical drawing board, because it depends on the existence of particles confined to two dimensions called anyons. These 'topological' particles are peculiarly impervious to environmental noise, in principle making them excellent qubits. Particles such as Majorana fermions that fulfil some of the requirements of anyons have been fabricated in certain solids, but whether they are useful for practical quantum computing is still debatable.

Extra benefits

As well as incredible computational power, quantum computers offer many other benefits. A key one is economy. In recent decades, we have done very well in cramming ever more transistors into classical computer chips. But the density of heat generated by constantly resetting these physical on-off switches now represents a fundamental barrier to further miniaturization. Quantum computation can sidestep that barrier.

This is because, by using the right manipulations, you can flick between quantum states such as the polarizations of photons without expending any heat. This is no blank cheque for low-power computing, however. Reading and writing information to a quantum memory entails making measurements akin to flipping a classical switch, and will still generate some heat.

Another benefit offered by quantum computing is the ability to tackle 'hard problems'. Computer scientists divide problems into 'easy' problems, where the computational resources

How quantum information processing took off

1981
Physicist Richard Feynman argues that modelling the correlations and interactions of particles in complex quantum physics problems can only be tackled by a universal quantum simulator that exploits those same properties.

1982
The **no cloning theorem** threatens hopes for quantum computing. It states that you cannot copy quantum bits, so there is no way to back up information. However, its advantage is that this makes intercepting data difficult – a boon for secure transmission of quantum information.

1985
David Deutsch at the University of Oxford shows how a universal quantum computer might, in theory, emulate classical logic gates and perform all the functions of quantum logic.

1984
Charles Bennett of IBM and Gilles Brassard of the University of Montreal in Canada develop BB84, the first recipe for secure encoding and transfer of information in quantum states.

1992
Superdense coding theory shows how a sender and receiver can communicate two classical bits of information by sharing only one entangled pair of quantum states.

1993
In fact, you do not need to transmit quantum states at all to exploit their power, as quantum teleportation protocols prove: it is sufficient to possess entangled quantum states and communicate using classical bits.

1995
US physicist Benjamin Schumacher coins the term **qubit** for a quantum bit.

1994
Shor's algorithm indicates how a quantum computer might factorize numbers faster than any classical computer.

1996
Grover's algorithm gives a recipe by which quantum computers can outperform classical computers in an extremely common task: finding an entry in an unsorted database.

1996
Quantum error correction theory finally overcomes the no-cloning problem. Quantum information cannot be copied – but it can be spread over many qubits. With this, the main theoretical tools for quantum information processing were in place. The challenge was then to create the technology!

needed to find a solution scale at some power of the number of variables involved, and 'hard' problems, where the resources needed increase in a much steeper exponential curve. Hard problems rapidly get beyond the reach of classical computers. The exponentially scaling power of a quantum computer could bring more firepower to bear – making the problems not exactly easy, but at least less hard.

A quantum speed-up is not a given, however: we first need a specific algorithm that tells us how to achieve it. For important, hard tasks such as factoring large numbers or searching a database, we already have recipes such as Shor's and Grover's algorithms (see 'Killer quantum computing apps' below). But for a mundane task such as listing all the entries in a database, the time or processing power required to solve the problem will always scale with the number of entries, and there will be no appreciable quantum speed-up.

These benefits, however, presume a quantum computer big enough to make a difference, and making one is easier said than done.

Interview: David Deutsch, quantum computing pioneer

David Deutsch is a professor of physics at the University of Oxford's Centre for Quantum Computation. In 1998 he received the Institute of Physics' Paul Dirac Prize and Medal, and in 2005 was awarded the Edge of Computation Science prize for work that extended the boundaries of the idea of computation. In 1985 Deutsch turned physics upside down by describing a universal quantum computer, pioneering the field of quantum information science. He explains how this relates to notions of truth and reality in our universe – and even outside it.

When you published your original paper on quantum computation, what was the general reaction?

People didn't take it on board as a new way of thinking. It took several years before the physics community began to work on quantum computation. There were a handful of people who saw the importance right away, but the field didn't begin until several years later.

What motivated you to begin working on quantum computation?

It was a desire to understand the foundations of quantum theory, of physics, of everything. The foundations of one field tend to overlap with the foundations of others. Quantum computation, for instance, has implications not only for the foundations of quantum theory but also the foundations of physics in general, and mathematics and philosophy.

How does quantum computation shed light on the existence of many worlds?

Say we decide to factorize a 10,000-digit integer, the product of two very large primes. That number cannot be expressed as a product of factors by any conceivable classical computer. Even if you took all the matter in the observable universe and turned it into a computer and then ran that computer for the age of the universe, it wouldn't come close to scratching the surface of factorizing that number. But a quantum computer could factorize that easily, in seconds or minutes. How can that happen?

Anyone who isn't a solipsist has to say that the answer was produced by some physical process. We know there isn't enough computing power in this universe to obtain the answer, so something more is going on than what we can directly see. At that point, logically, we have already accepted the many-worlds structure. The way the quantum computer works is: the universe differentiates itself into multiple universes and each one performs a different sub-computation. The number of sub-computations is vastly more than the number of atoms in the visible universe. Then they pool their results to get the answer. Anyone who denies the existence of parallel universes has to explain how the factorization process works.

Want one today?

Can't wait to get your hands on a shiny new quantum computer? The good news is that you can buy one today, if you have at least $10 million to spare.

D-Wave Systems of Burnaby, Canada, is the newcomer breaking in on the quantum action. It counts Google and NASA among its customers. Its flagship model, launched in January 2017, known as D-Wave 2000Q, contains a lattice of tiny superconducting loops of the metal niobium, each of which is one qubit. It has up to 2,048 qubits, and uses 128,000 Josephson junctions (see Figure 5.3). But be warned: this is no quantum laptop. The whizzy-looking black box housing it, along with its supporting cryogenic system and supercomputer interface, fills a room. Perhaps surprisingly, then, it runs on just 25 kilowatts, a fraction of the power devoured by the world's fastest supercomputers.

FIGURE 5.3 A processor from the D-Wave computer

A full 2,000-qubit performance would far outdo that of its rivals, and their tests show that this new D-Wave machine outperformed a classical one by 1,000 to 10,000 times, in terms of pure computation time. But tests on its predecessor, the D-Wave 2X, have proved inconclusive as to whether it really outperforms ordinary computers.

In August 2015 D-Wave systems claimed that its D-Wave 2X is up to 15 times faster than regular PCs. It put the machine through its paces with a series of tests based on solving random organization problems, such as picking the best football team from a squad of players, all with different abilities and who work better or worse in different pairs. Compared to specialized optimization software running on an ordinary PC, the 2X found an answer between two and 15 times as quickly. But critics say that this was not a fair comparison.

The D-Wave 2X has only one application: an optimization algorithm that searches for the best solution to a given problem. However, that's enough for D-Wave's first two customers.

Google is using it in machine learning, and Lockheed Martin is using the machine to find out where its aircraft software might go wrong.

Killer quantum computing apps

There is no shortage of potential uses for quantum computers. Here is a rundown.

Ultrasecure encoding

This quantum information technology is already up and running. Various small-scale quantum cryptographic systems for secure information transfer, typically using polarized photons as their qubits, have been implemented by labs and companies such as Toshiba, Hewlett Packard, IBM and Mitsubishi. In October 2007 a quantum cryptography system developed by Nicolas Gisin and colleagues at the University of Geneva in Switzerland was used to transmit votes securely between the city's central polling station and the counting office during the Swiss federal elections. A similar trial system developed by the researchers' company, ID Quantique, was used to transmit data securely during the 2010 Football World Cup in South Africa.

The distance through which quantum states can be transmitted through fibre-optic cables is limited to tens of kilometres owing to random diffusion. One promising way to get around this is akin to error-correction protocols devised for quantum computers: to spread information over more than one qubit. But this might pose a security risk by giving more information for an eavesdropper to hack.

Transmission via air is an alternative. The world record in faithfully teleporting a qubit of information, held by Anton Zeilinger and his colleagues at the University of Vienna, Austria, is over a distance of 143 kilometres between the Canary islands of La Palma and Tenerife. This indicates that delicate quantum states can be transmitted significant distances through air without being disturbed – and suggests that a worldwide secure quantum network using satellites is a distinct possibility.

In August 2016 China launched the world's first quantum communications satellite to test out technology that could one day be part of an unhackable network. It will use photons to test quantum key distribution.

Quantum simulation

Richard Feynman's original motivation for thinking about quantum computers in 1981 was that they should be more effective than classical computers at simulating quantum systems – including themselves. This sounds a little underwhelming, but many of science's thorniest practical problems, such as what makes superconductors superconduct or magnets magnetic, are difficult or impossible to solve with classical computers.

Quantum information theorists have already developed intricate algorithms for approximating complex, many-bodied quantum systems, anticipating the arrival of quantum computers powerful enough to deal with them. The beauty is that such simulators would not be limited to existing physics: we could also use them to glean insights into phenomena not yet seen. Quantum simulations might tell us, say, where best to look in nature for Majorana particles, for example in complex many-bodied superconductor states. Since these particles, thought to be their own antiparticles, have properties that could make them ideally suited to making robust quantum memories, this raises the intriguing possibility of using quantum computers to suggest more powerful quantum computers.

Metrology

Making precise measurements is a potentially highly significant application of quantum computers. When we record sensitive measurements of physical quantities, such as intervals in time or distances in space, the effects of classical noise mean that the best statistical accuracy we can achieve increases with the square root of the number of bits used to make the recording.

Quantum uncertainty, meanwhile, is determined by the Heisenberg uncertainty principle and improves much more rapidly simply with the number of measurements made. By encoding distances and time intervals using quantum information – probing them using polarized laser photons, for example – much greater accuracies can be achieved.

This principle is already being applied in giant 'interferometers', which used the disturbances to kilometre-long laser beams to detect the elusive gravitational waves predicted by Einstein's relativity, such as the LIGO detectors in Livingston, Louisiana, and Hanford, Washington State. In these cases, we can think of gravity as noise that disturbs qubits – the qubits being the position and momentum of laser photons. By measuring this disturbance, we can estimate the waves' strength.

Are we nearly there yet?

No overview of quantum computing would be complete without an attempt to answer the $64,000 (or possibly much more) question: are we likely to see working quantum computers in our homes, offices and hands any time soon? That depends largely on finding a medium that can encode and process a number of qubits beyond the 10 or 20 that current technologies can handle. But getting up to the few hundred qubits needed to outperform classical computers is largely a technological issue. Within a couple of decades, given improvements in cooling and trapping, as well as coupling with light, existing technologies of trapped ions and cold atoms may well be made stable enough in large enough quantities to achieve meaningful quantum computation.

The first large-scale quantum computers are likely to be just that: large-scale. Investment in this field has picked up in recent years, and even sceptics are now talking of large-scale quantum computation as an inevitable development, perhaps coming to fruition within the next five to ten years. These systems will probably require lasers for qubit manipulation and need supercooling, so are unlikely to appear in our homes. But if the future of much computing is in centralized clouds, perhaps this need not be a problem.

When it comes to anything smaller, the obvious problem is entanglement, which is a fragile good at the best of times and becomes harder and harder to maintain as the quantum system grows. It would aid the progress of quantum computing if our assumption that entangled states are an essential, central feature turned out to be wrong. This intriguing possibility was raised in 1998 with the development of 'single-qubit' algorithms. These can solve a large class of problems, including Shor's factorization algorithm, without the need for many entangled qubits. That would be a remarkable trick if it could be pulled off in practice – although Grover's all-important database-search algorithm might still not be implementable in this way.

Some people believe that the fragility of quantum systems will never allow us to implement quantum computation in the sort of large, noisy, warm and wet environments in which we humans work. But we can draw hope from recent evidence that living systems – such as photosynthesis in bacteria and retinal systems for magnetic navigation in birds – might be employing some simple quantum information processing to improve their own efficiency. (See Chapter 6 for more on this.)

If we can learn such secrets, a quantum computer on every desktop and in the palm of every hand no longer seems so fanciful an idea.

Quantum communications

Present-day cryptographic systems are in a precarious state. The security of all of our online purchases, bank transactions and personas rely on a single shaky assumption: that certain mathematical operations are hard to do. The best known of our modern encryption systems is called RSA. To encode data, it builds a key from two very large prime numbers. These are kept secret, but their product – a number thousands of binary digits long – is public knowledge. Data can be encoded using this public key, but only those with knowledge of the original numbers can decrypt it. RSA's security relies on the fact that there is no known shortcut to find the two starting numbers. The only ways to do it are almost interminable processes, such as trying all the possibilities one by one.

Or so we hope. At present, no conventional computer has the power to quickly solve these 'brute force' calculations, but that could change, not least if large-scale truly quantum computers are made to work. One way to reinvigorate our privacy is to employ quantum cryptography (see Figure 5.4). This promises the ability to create keys that are entirely random, entirely unpredictable and totally inaccessible to spies.

FIGURE 5.4 The eye of an observer reflected in a mirror in quantum cryptography apparatus. Cryptography in this case refers to the encoding of data so that only specified targets can access it. If the sender and the receiver of the data share entanglement, then any data is only readable by them alone. This is more secure than existing encryption methods, most of which rely on mathematical algorithms.

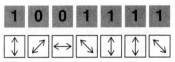

1. To distribute a secret key, Alice first sends Bob a random sequence of bits represented by photons polarized in one of four states.

2. Bob has two detectors which he uses randomly. One detects ↗ ↘, the other detects ↕ ↔, meaning he detects only some of these photons correctly. Alice confirms which bits match her initial choices.

3. They then discard the incorrect ones and use the remaining bits as their secret key.

4. This is used to encrypt a message sent over a non-quantum channel, such as the Internet.

FIGURE 5.5 Unbeatable security: how quantum keys will keep messages safe

Quantum cryptography hinges on the rules that govern particles like photons or electrons. Their properties, including polarization, for instance, take multiple values at once, only snapping into sharp definition when measured (see Figure 5.5). Use these properties as a basis for encryption and you preclude any attempt to peek at your key: that would change the result of the measurement, in effect destroying the key's tamper-proof seal.

Current systems use a protocol where the person transmitting the key, usually referred to as Alice, releases a polarized photon and makes a measurement on it before sending it. Her listening partner, usually referred to as Bob, chooses a particular way to make a measurement of that polarization, and then he and Alice use an unencrypted channel to compare the sort of measurements they did. This allows them to create one digit of a private key for use in encrypting messages. To build the entire key, Alice and Bob simply repeat the process.

The technique has already been used to protect hospital data, financial transactions and voting data in a Swiss federal election. The technology is rather expensive, however, so now the challenge is to develop cheap quantum communication devices. A prototype of a quantum communication transmitter chip has already been developed at the University of Bristol, and one day such devices could sit inside your Internet router or mobile phone to bring secure communication to the masses.

Interview: A quantum algorithm to unravel online data encryption?

Peter Shor is a professor of applied mathematics at the Massachusetts Institute of Technology. His quantum algorithm could crack the encryptions that protect our data online – but only if a powerful enough quantum computer is built. Here he explains why he devised the algorithm for a quantum computer that could unravel our online data encryption.

Internet security relies on the fact that our computers can't break its cryptosystems. But the quantum algorithm you devised has the potential to do just that. Why create it?

My motivation was to see what you can do with a quantum computer. An earlier quantum algorithm worked by using periodicity – the tendency of some number sequences to regularly repeat. This is related to factoring, or finding which smaller numbers big numbers are divisible by, so I thought a quantum computer might be able to factor large numbers. As Internet cryptosystems rely on the fact that current computers cannot factorize big numbers, I figured a powerful enough quantum computer could break these systems.

Did you worry about the implications when you finished 'Shor's algorithm' in 1994?

I felt great having discovered something nobody else knew. If I hadn't done it, someone else would have, eventually. At that time quantum computers were completely hypothetical and I didn't really think one could be built. We are pretty safe for five or ten years, probably more.

Quantum cryptography can't be broken by factorization. Could it one day replace standard cryptosystems?

For short distances it wouldn't be too hard to build a quantum key distribution network to encrypt data. Over longer distances, you would need quantum repeaters every 50 kilometres or so on the fibre-optic network, as it's difficult to maintain a quantum state over long distances. Even if they are cheap by then, it's a lot of investment.

How much harder is it to write an algorithm for a quantum computer?

It's much harder. Quantum computers rely on a form of interference – basically the same phenomenon as light-wave interference, in a more mathematical setting. Computational paths to the right answer need to interfere constructively, while those to the wrong answer should interfere destructively.

Why write quantum algorithms when we don't yet have proper hardware to run them?

The more things that you can do with quantum computers, the more important it is to build them. For example, you should be able to use them to better design drugs using quantum effects that predict the molecules' chemistry. Right now, drug companies use conventional software to simulate those effects, but you might do better if you could use an actual quantum computer.

Noise: the key to quantum technologies?

Quantum devices are proving extremely difficult to build because they must operate under noise-free conditions. Controlling entanglement, however, is a frustrating goal. Collisions with air molecules, stray electromagnetic signals, heat and many other factors create vibrations, or noise, which rapidly degrade this quantum feature. A quantum algorithm that corrects for noise can preserve entanglement long enough to compute accurate results, but so far no general-purpose quantum computer has been built using more than a few qubits. A large-scale computer based on pure entanglement is unlikely to become a reality in the next decade.

What if we could build quantum devices that tolerate noise – or even exploit it? This may be possible, thanks to an obscure property of the quantum world called discord – a hot but controversial topic. Discord was first discovered in the early 2000s by three groups working independently in the UK, the United

States and Poland. At its basis is the realization that quantum is not a yes/no option. A system can be fully quantum, and hence criss-crossed with entangled connections. But it can also be only partially quantum, lacking entangled connections but possessing other features of quantum theory. In essence, discord measures this quantumness, encompassing both entanglement and what was once thought of as unwanted noise. It is ubiquitous in quantum systems.

For seven years, discord remained a niche topic and its practical importance was unclear. However, interest in discord increased five years ago when evidence began to accumulate that it can add 'quantum power' to a system even when entanglement is absent. Previously, it had been assumed that entanglement was a critical requirement.

A key development came in 2008 when researchers at the University of New Mexico took a fresh look at a simplified model of a quantum computer called DQC1. They found that, as more qubits are added, the computer continues to work efficiently even as the number of data entries grows exponentially – something impossible to achieve with digital computers. Interestingly, this large improvement was achieved without a large increase in the amount of entanglement, leading to the conclusion that discord had enabled it. Remarkably, the DQC1 model works using only one qubit protected from noise, while all its remaining qubits are fully noisy. It shows us that a whole lot of noise need not be a nuisance, but rather a resource when put together with just a little bit of clean signal. Other recent research shows that a quantum computer that generates no discord cannot be more powerful than a classical one, in many instances.

Discord appears to play a useful role in quantum sensors, too – a potential way to boost sensor precision while using less energy. These could be used to probe fragile biological specimens that are damaged by light exposure, for example. Research shows that in some types of noisy quantum sensors, where entanglement is unprotected from noise, a boost in quantum power is possible using techniques that exploit discord.

Another exciting finding is that there is a direct correlation between discord and the precision of quantum sensors. The DQC1 quantum computer has also been applied to quantum sensors, demonstrating how discord enables quantum-enhanced measurement precision.

Discord, however, is a controversial topic, and some researchers are sceptical of its significance. This is largely because it has taken many years to develop a clear picture of discord as an important and usable physical quantity. But this view is changing as more applications that utilize discord are being uncovered.

Consuming discord

One of the most exciting discoveries came in 2012 when a team led by Mile Gu of the Centre for Quantum Technologies and Ping Koy Lam of the Australian National University in Canberra showed a convincing link between quantum power and discord. They demonstrated that the improvement in the amount of information about a secret encoded message that can be extracted using a true quantum machine is equal to the discord consumed in the process.

The once-obscure line of discord research is rapidly becoming mainstream, now that there is clear evidence that noisy quantum devices will provide stepping stones to powerful quantum technologies. Discord might even play a role in our understanding of the quantum-to-classical transition that explains the emergence of our everyday experience of the real world, as well as other foundational issues in physics. Quantum-enhanced measurements are likely one day to be used in state-of-the art sensors for geophysical exploration and other fields.

Record breakers

Quantum teleportation
The long-distance record for quantum teleportation between two locations on Earth is 143 kilometres. It was set in 2012 by a group led by Anton Zeilinger at the University of Vienna. A new world record for quantum teleportation over fibre-optic networks was set in September 2016, when two independent teams transferred quantum information over a distance of 6.2 kilometres.

Superposition
The biggest object to exist in two quantum states at the same time is a cloud of 10,000 rubidium atoms.

Transmission through space
The first quantum transmission to go via space was achieved in 2014, when photons in four different quantum states (the minimum required for quantum cryptography) were sent to space and back, bouncing back from five satellites located up to 2,600 kilometres away.

6
Quantum biology

We tend to think that the interaction between quantum physics and biology stops with Schrödinger's cat (not that Erwin Schrödinger intended his unfortunate feline to be anything more than a metaphor). Indeed, when he wrote his 1944 book What is Life? *he speculated that living organisms would do everything they could to block out the fuzziness of quantum physics. But is that really still the case?*

Has life harnessed the power of quantum mechanics?

Quantum mechanics seems very odd to us because we don't see these phenomena in the visible world. Instead, they are normally confined to the tiniest constituents of matter that lie beyond our senses, such as electrons or atoms. And when trillions of microscopic particles come together inside big bulky objects, all the quantum weirdness stuff gets somewhat washed away by the consequent molecular vibrations, or noise.

This is why, when scientists want to investigate quantum phenomena, they usually have to do so in rarefied laboratory environments. They must cool everything down to close to absolute zero, pump out all the air and shield their experiments from any extraneous vibrations. Only then can they detect the delicate quantum phenomena.

Life is hot, messy and complicated. It happens inside living cells full of jostling particles generating a cacophony of molecular noise. It would seem to be the last place where you would expect to find quantum mechanics. Yet in recent years, it has become increasingly clear that quantum behaviour can survive inside living cells. It might explain photosynthesis, the action of enzymes, the way birds navigate and even how DNA works. Life seems to have evolved to exploit the weirdness of quantum mechanics to keep us alive.

Let there be light: photosynthesis

Photosynthesis is one of life's most important chemical reactions. It uses energy from sunlight to make biological building blocks, but its extraordinary efficiency is hard to explain.

The first step in photosynthesis is the capture of a photon of light by an electron in the outer shell of a magnesium atom inside a molecule of chlorophyll pigment. The extra energy causes the electron to vibrate, forming something called an exciton.

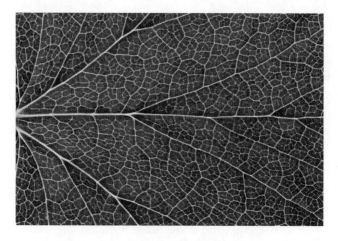

FIGURE 6.1 Do leaves use quantum mechanics to capture light?

The next step is the transport of the exciton to a reaction centre where the captured light energy can be turned into chemical energy. This transfer has to be very quick, otherwise the light energy will be lost. But, to find the reaction centre, the exciton has to travel through a forest of pigment molecules where its energy is most likely to get lost. Yet measurements show that the exciton transport has the highest efficiency of any energy transport reaction, close to 100 per cent under optimal conditions. This level of efficiency is difficult to explain using the laws of classical physics.

What is going on? In 2007, experiments led by Greg Engel, now at the University of Chicago, shone laser light at the photosynthetic system of bacteria. He got a kind of light echo back. But what was odd was that the echo was returned as a beating wave. This beating is a signature of quantum behaviour. It showed that the exciton was not taking one particular route through the photosystem: it was taking all possible routes, travelling to the reaction centre as a quantum wave. This was the first direct evidence that, at its heart, photosynthesis is a quantum mechanical process. Since then, quantum coherence has been found in many bacterial and plant photosystems and appears to be a fundamental feature of how plants and microbes capture light energy. Remarkably, this fragile state survives even at ambient temperatures, a feature that provokes both interest and envy from quantum computer designers who usually have to do their computations at temperatures close to absolute zero.

Could quantum photosynthesis lead to the creation of better solar cells?

That's the hope. Greg Engel of the University of Chicago, who first discovered the quantum beat in photosynthesis, has been making synthetic pigment molecules that have the same coherence properties as the pigment molecules found in photosynthetic complexes of cells. The aim is eventually to make solar cells that can transport energy as efficiently as happens in nature. But this will take many years to come to fruition and, before this can be achieved, we need to find out more about exactly how life manages to maintain coherence for so long.

Quantum compass: magnetoreception

The most famous proposed example of quantum biology comes from the navigation system of the European robin (see Figure 6.2). In winter this bird flies south to escape the freezing north; and, to help it navigate, it uses a compass. European robins have a special type of inbuilt compass that can measure the angle of the Earth's magnetic field against the surface of the Earth.

FIGURE 6.2 A robin (*Erithacus rubecula*) in flight; the migration of robins might be aided by a quantum compass.

How does this 'inclination' compass work? In 2000 Thorsten Ritz of the University of California came up with the idea that it might depend on a peculiar feature of quantum entanglement. When two entangled particles are electrically charged, they can detect the angle between them and the Earth's magnetic field. Ritz proposed a model for the avian magnetoreception, in which light generates a pair of entangled particles in the eye of the robin that become a compass, able to detect the angle of inclination of the Earth's magnetic field. One of the predictions of this theory was that the quantum-entangled compass should be disrupted by high-frequency radio waves – and tests demonstrated exactly this.

This does not prove that the robin's compass is quantum mechanical, but so far no one has come up with an alternative explanation of the results of Ritz's experiments.

Making it happen: enzymes

Enzymes are the engines of life. They manipulate electrons, protons, atoms and molecules to make every biomolecule in our bodies. And they are incredible catalysts that can speed up chemical reactions by a factor of 10^{20}. If a similar boost were provided to your walking pace, you could hop to another galaxy.

Just as for photosynthesis, it is hard to account for this astonishing acceleration of chemical reaction by the classical laws alone. However, it is now becoming clear that enzymes gain their huge chemical acceleration by manipulating the quantum mechanical nature of matter, employing a process called quantum tunnelling. This is where a particle can travel through a seemingly impenetrable barrier using its wave properties, essentially dematerializing from one point in space and materializing in another, without visiting any of the in-between places.

Back in the 1970s, research demonstrated that enzymes involved in respiration transfer electrons by quantum tunnelling. Electrons are small so this was not too surprising, but more recent research has shown that enzymes also promote the hopping of much more massive protons from one atom to another by quantum tunnelling. Our bodies may have been constructed in the quantum world.

Something odorous: smell

Our sense of smell is extraordinarily powerful. It can detect tiny amounts of chemicals, even single molecules. And it is extraordinarily specific, allowing us to differentiate between thousands of smells. But how does it work?

The conventional explanation is that it works by a kind of lock-and-key mechanism. An odorant molecule floats through the air to be captured by an olfactory receptor protein in your nose.

The odorant and receptor molecules are thought to fit together like a key in a lock. But there are many problems with the theory, such as the fact that similarly shaped odorant molecules – essentially the same set of keys – often smell very different. Conversely, very different molecules often smell the same. The chemist Malcolm Dyson came up with an alternative theory in the 1920s, when he proposed that it was not the shape of odorants that is detected, but their molecular vibrations.

But nobody knew how a molecular vibration detector would work in the nose. Then, in the 1990s, the biochemist Luca Turin proposed that the nose employs a vibration detector that utilizes quantum tunnelling. The theory predicted that odorants made with different isotopes of elements should smell different. Experiments reported in 2013 demonstrated that fruit flies could indeed distinguish isotopic odorants, as predicted by the theory.

Written in code: DNA mutations

Could evolution be aided by quantum mechanics? This is something that biologist Johnjoe McFadden and physicist Jim Al Khalili, both of the University of Surrey, UK, have been interested in for some years.

Evolution works through the generation of variation in DNA sequences by mutation. Essentially, the wrong DNA letter is incorporated into DNA during its replication. Natural selection then takes over to select for beneficial mutations.

At the heart of DNA's double helical structure are hydrogen bonds that hold the helix together. These bonds are based on single protons, shared between two positions on the double helix. This means that DNA, life's blueprint, is written in quantum mechanical letters. In the 1970s the Swedish physicist

Per-Olov Löwdin proposed that quantum tunnelling of coding protons can promote mutations in DNA. Many studies have shown that this is theoretically feasible; but, thus far, nobody has nailed the mechanism in an experiment.

Could quantum biology lead to a new definition of life?

In their book *Life on the Edge: The Coming of Age of Quantum Biology* (2014), Jim Al-Khalili and Johnjoe McFadden make the claim that the ability to maintain coherent quantum states for a significant length of time is fundamental to life and that a definition of life should include the quantum trickery that cells carry out. They argue that this is what makes life different from everything else. McFadden admits that this is a speculative claim that doesn't yet have evidence to back it up, but he hopes that there will eventually be a way of testing it through advances in synthetic biology, when new life forms are created artificially.

Where does this leave us?

Random molecular noise normally destroys quantum effects in inanimate systems. How, then, do quantum effects survive in hot, wet and molecularly noisy, living cells? One of the most surprising and intriguing features of quantum biology is that life seems to have found ways of using molecular noise to maintain, rather than destroy, quantum coherence. Indeed, this may be one of the fundamental attributes of life. Molecular noise may allow life to navigate the edge between the quantum and classical worlds.

In search of the quantum brain

Does the brain use quantum mechanics? On one level, the answer is yes. The brain is composed of atoms, and atoms follow the laws of quantum physics. But what about the related question of whether the strange properties of quantum objects – being in two places at once, seeming instantly to influence each other over distance and so on – could explain still-perplexing aspects of human cognition? That, it turns out, is a very contentious question indeed.

The most basic objection comes from Occam's Razor, the principle that says the simplest explanation is usually the best. In this view, current non-quantum ideas of the brain's workings are doing just fine and we have no need to invoke quantum physics to explain cognition. However, theoretical physicist Matthew Fisher of the University of California, Santa Barbara, is less sure, pointing out that current ideas about memories are far from watertight – for example, that they are stored in the architecture of neuron networks or in the junctions between neurons. Why not see whether there are better quantum explanations?

It is perhaps because we've been here before. In 1989 Oxford mathematician Roger Penrose proposed that no standard, classical model of computing would ever explain how the brain produces thought and conscious experience. The suggestion intrigued a lot of people, not least an Arizona-based anaesthetist named Stuart Hameroff, who suggested a specific way for quantum effects to get involved.

Simultaneous answers

The crux of the idea was that microtubules – protein tubes that make up neurons' support structure – exploit quantum effects to exist in 'superpositions' of two different shapes at once. Each of these shapes amounts to a bit of classical information, so this shape-shifting quantum bit, or qubit, can store twice as much information as its classical counterpart.

Add entanglement to the mix – a quantum feature that allows qubit states to remain intertwined even when not in contact – and you rapidly build a quantum computer that can manipulate and store information far more efficiently than any classical computer. In fact, Penrose suggested, the way such a computer can arrive at many answers simultaneously, and combine those

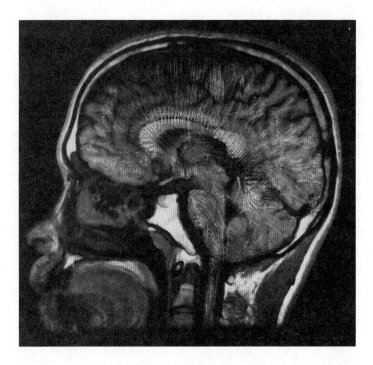

FIGURE 6.3 The way a quantum computer works has been compared to the way the brain works.

answers in different ways, would be just the thing to explain the brain's peculiar genius (see Figure 6.3).

Penrose and Hameroff collaborated on the idea, and they and others considered it to be a sensible proposal for a while. But holes soon began to appear. From a physicist's perspective, the most fundamental problem was coherence time. Superposition and entanglement are both extremely fragile phenomena. Think of a human pyramid of performers crossing a high wire on a unicycle and you get the idea. The slightest disturbance and their grip slips. In the case of a quantum system, it will 'decohere' to a standard classical state if disturbed by heat, a mechanical vibration or anything else. The information stored in the quantum states is generally lost to the surrounding environment.

This problem has hampered attempts over the past two decades by physicists, Fisher included, to engineer a quantum computer of any significant size. Even in cryogenically cooled and mechanically isolated conditions, it is difficult to keep qubit networks coherent for long enough to do anything beyond the capabilities of classical computers. In the warm, wet brain, with its soup of jiggling, jostling molecules, it becomes almost impossible. Neurons hold information for microseconds at a time or more while processing it, but calculations suggest that the microtubule superpositions would last only between 10^{-10} and 10^{-13} seconds.

Evolution

Fisher shared that scepticism, which was shared by many others. But equally, he thought, would it not be odd if evolution had not worked that out? Life has had billions of years to 'discover' quantum mechanics, and its exquisite molecular apparatus gives it the means to exploit it. Even if electrical impulses among neurons within the brain – something well described by classical physics – are the immediate basis of thought and memory, a hidden quantum layer might determine, in part, how those neurons correlate and fire.

Fisher's personal interest in the subject began in a rather roundabout way, while he was wondering about the persistence of mental illness among people close to him, as well as the efficacy of the drugs used to treat them. The initial focus of Fisher's interest was lithium, an ingredient of many mood-stabilizing drugs. As he combed the scientific literature, he happened across one particular report from 1986 that gave him pause for thought. It described an experiment in which rats were fed one of the two stable isotopes of lithium: lithium-6 and lithium-7. When it came to grooming, nursing of pups, nest building, feeding and several other measures, those fed lithium-6 were enormously more active than control groups or those fed lithium-7.

It was this paper that led Fisher to think it might be time to address the whole troubling problem of quantum cognition once again. All atomic nuclei, like the fundamental particles that make them up, have a quantum-mechanical property called spin. Crudely, spin quantifies how much a nucleus 'feels' electric and magnetic fields: the higher the spin, the greater the interaction. A nucleus with the very lowest possible spin value, 1/2, feels virtually no interaction with electric fields and only a very small magnetic interaction. In an environment such as the brain, where electric fields abound, nuclei with a spin of 1/2 would be peculiarly isolated from disturbance.

In a spin

Spin-1/2 nuclei are not common in nature, but here's the thing. The spin value of lithium-6 is 1, but in the sort of chemical environment found in the brain, a water-based salt solution, the presence of the water's extra protons is known to make it act like a spin-1/2 nucleus. Experiments as long ago as the 1970s had noted that lithium-6 nuclei could hold their spin steady for as long as five minutes. If there is an element of quantum control to the brain's computation, Fisher reasoned, lithium's calming effects might be down to the incorporation of these peculiarly coherent nuclei into the brain's chemistry.

This is not the only thing. Lithium-6 does not occur naturally in the brain, but one nucleus with a spin-1/2 does, and it is an active participant in many biochemical reactions: phosphorus.

After exhaustive calculations of the coherence times of various phosphorus-based molecules in biological settings, Fisher went public with a candidate qubit. It is a calcium phosphate structure known as a Posner molecule or cluster. It was identified in bone mineral in 1975 and has also been seen floating around when simulated body fluids – that is, water with added biological molecules and mineral salts – are concocted in the lab.

When Fisher estimated the coherence time for these molecules, it came out as a colossal 10^5 seconds – a whole day. He has also identified at least one chemical reaction in the brain that he thinks would naturally manufacture entangled, coherent states between nuclear spins within Posner molecules. It is a process involved in calcium absorption and fat metabolism that uses an enzyme called pyrophosphatase. This enzyme breaks down structures made of two interlinked phosphate ions, producing two single ions. Theoretically, at least, the nuclear spins in these two ions should be quantum entangled. Release them into the fluid surrounding the cells, and they can combine with calcium ions to form Posner molecules.

If this is all correct, the brain's extracellular fluid could be awash with complex clusters of highly entangled Posner molecules. Once inside the neurons, these molecules could begin to alter the way the cells signal and respond, starting to form thoughts and memories (see Figure 6.4).

Fisher published the details of his proposal in *Annals of Physics* in 2015. Much of it, he admits, is highly speculative. The first test will be whether Posner molecules exist in real extracellular fluids. If they do, can they be entangled? This is just one of the many controversial ideas in this field.

Consciousness

Penrose is – perhaps predictably – excited by the story so far. Penrose still favours his microtubule hypothesis, however, seeing the new proposal as a mere add-on that allows for lasting memory. For Penrose, consciousness has to do with gravity acting on quantum states and thus causing them to decohere; microtubules are more massive than nuclei, and thus more likely to be the cause of this interaction, he says.

The questions keep coming. Does a bang on the head induce memory loss because it causes decoherence? Is nuclear spin the reason you can change brain states with transcranial magnetic stimulation, which fires a magnetic field across the brain?.

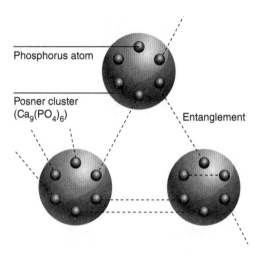

Phosphorus atom

Posner cluster
$(Ca_9(PO_4)_6)$

Entanglement

Change the spin state of one entangled phosphorus
atom and the state of its entangled partner changes
too – regardless of how far apart they are.

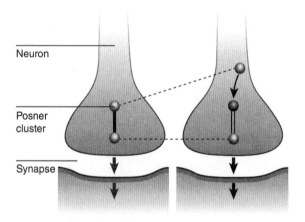

Neuron

Posner
cluster

Synapse

Entangled Posner clusters involved in chemical
signalling in one brain neuron could induce similar
reactions in another neuron.

FIGURE 6.4 Posner clusters thought to be found in the brain contain six
phosphorus atoms whose nuclear spin states can be quantum entangled –
perhaps influencing how we think and remember.

More fuzzy logic

Schrödinger suspected that the human mind might be beyond human understanding – but quantum mechanics could help us understand the way we think.

Human thinking, as many of us know, often fails to respect the principles of classical logic. We make systematic errors when reasoning with probabilities, for example. But recent research has shown that these errors actually make sense within a wider logic based on quantum mathematics. The same logic also seems to fit naturally with how people link concepts together, often on the basis of loose associations and blurred boundaries. That means search algorithms based on quantum logic could uncover meanings in masses of text more efficiently than classical algorithms.

It may sound preposterous to imagine that the mathematics of quantum theory has something to say about the nature of human thinking. This is not to say there is necessarily anything quantum going on in the brain – only that 'quantum' mathematics is not really owned by physics at all, and turns out to be better than classical mathematics at capturing the fuzzy and flexible ways that humans use ideas.

One example of the way that human thinking defies classical logic was demonstrated in the early 1990s by psychologists Amos Tversky and Eldar Shafir, of Princeton University, in a two-stage gambling experiment. They showed that even though the outcome of the second stage does not depend on the first stage, a participant's decision to enter the second gamble is influenced by whether or not they are told how they did in the first gamble. But this outcome defies classical logic.

This and other experiments demonstrate that people are not logical, at least by classical standards. But quantum theory offers richer logical possibilities and may provide a better framework for modelling human decision-making. The seemingly illogical decisions demonstrated by Tversky and Shafir, for example, can be explained using models based on quantum interference.

Another challenge is to get computers to find meaning in data in much the same way that people do. If you want to research a topic such as the 'story of rock' with geophysics and rock formation in mind, you do not want a search engine to give you millions of pages on rock music. One approach would be to include '-songs' in your search terms in order to remove any pages that mention 'songs'. This is called negation and is based on classical logic. While it would be an improvement, you would still find plenty of pages about rock music that just don't happen to mention the word 'songs'. Research by computer scientist Dominic Widdows has found that a negation based on quantum logic works much better.

This fits with the views of some psychologists, who argue that strict classical logic plays only a small part in the human mind, where much of our thinking operates largely on an unconscious level where thought follows a less restrictive logic and forms loose associations between concepts.

7
In search of reality

For all its phenomenal accuracy, quantum theory as it stands remains deeply unsatisfying. It was mainly conjured up to explain weird experimental results, and its mathematical foundations are shaky and incomplete. Is there a way to solidify them, or do we need to tear down the edifice altogether in favour of something new?

Probing the weirdness

Quantum theory is our best theory of basic physics, with many experimental and practical successes to its name (see Chapters 4 and 5). But for many, there are too many issues with the theory to think it anything more than an approximation of reality. One of the most perplexing questions left open by quantum theory is that of scale: where does the quantum world end and the classical world begin? Thanks to a new generation of experiments now being developed, we could at last have the sensitivity to get an answer.

Take the strange phenomenon of superposition. Its most famous manifestation is in the double-slit experiment, where a photon is observed to go through two slits at the same time and somehow interfere with itself (see Chapter 2). In other words, as long as nobody is watching, the photon exists in two different places at once.

The situation is analogous to that of Schrödinger's cat. Schrödinger came up with this bizarre scenario to show that there was something wrong with quantum theory. There is no way, he said, that something as non-quantum as a cat can be in a superposition of alive and dead – whether it is being observed or not. But since then researchers have demonstrated that carbon-70 molecules can go through two slits at once, too. Although these ball-shaped molecules aren't quite as substantial as cats, they can nonetheless be seen through a microscope.

Such experiments have been extremely useful in teaching us about what constitutes a measurement. They have shown, for example, that if conditions allow an observer to infer which slit the photon went through – if, say, there were stray photons in the apparatus that could bounce off the test photon and thus give away its position – the superposition will disappear.

This destruction, or collapse, of the superposition is known as decoherence.

Exploring when decoherence occurs has allowed us to find out more about what makes the quantum world tick. However, there is still an enormous amount we don't know. And here we are running up against a difficult logistical problem.

Quantum machines

Pushing at the boundary between the quantum world and that of classical physics means using ever larger molecules to see where decoherence destroys superposition. But the bigger the molecule, the harder it is to control outside forces and stop them from disrupting the molecule's delicate quantum state. For large molecules, uncontrolled decoherence effects rule, spoiling the very effect you want to measure.

This is where the quantum machines come in. At the moment, they do not look like much. The biggest of them is little more than a sliver of aluminium about 50 micrometres in length. It functions as an oscillator, something like a quantum tuning fork. The key is its mass. Even the relatively large clusters of carbon atoms being used today are lightweights compared with the mass that quantum machines will have (see Figure 7.1).

This is useful because the mass of a quantum object plays an important role in several alternative explanations of how the quantum world works. In 2003 Oxford mathematician Roger Penrose suggested that gravity might cause superposition collapse. Penrose reckons we will eventually be forced to combine Schrödinger's equation describing quantum particles, an understanding of the measurement process, and the principles of Einstein's theory of gravity into one theory, and that each of these three aspects of reality will then be seen as nothing

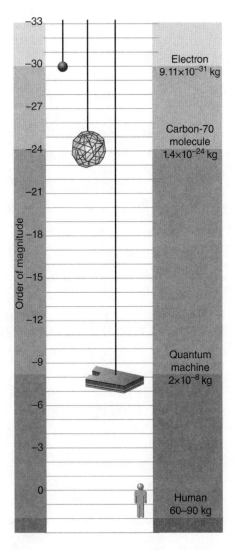

FIGURE 7.1 Mass matters: quantum machines are orders of magnitude more massive than anything that has demonstrated quantum behaviour before, such as electrons and molecules.

more than an approximation to a deeper fundamental truth (see Chapter 8).

Testing such ideas, however, will require quantum machines of almost incomprehensible sensitivity. The necessary apparatus involves mirrors 10 micrometres across that weigh just a few trillionths of a kilogram, with mechanisms capable of registering less than a billionth of a millimetre. Such quantum machines would be a move towards observing genuinely macroscopic objects in two places at once. If we can get there, it would be the mechanical analogy of a cat being alive and dead. There is still some way to go before we are quite ready to resolve that paradox.

The mystery of matter deepens

At the heart of quantum theory is the idea that light can exist as a particle and a wave at the same time. But is that the way things really are, or just the easiest way for our classical brains to understand it?

For Niels Bohr, the great Danish pioneer of quantum physics, the 'central mystery' of this new field was working out why photons sometimes behave as waves and at other times as particles. He called this principle complementarity: quantum objects such as photons simply have complementary properties – being a wave, being a particle – that can be observed singly but never together. And what determines which guise an object adopts? We do. Look for a particle and you'll see a particle. Look for a wave and that's what you'll see (see Figure 7.2).

The idea that physical reality depends on an observer's whim bothered Einstein no end. Does the moon's existence depend on observers too, he once asked. 'No reasonable definition of reality could be expected to permit this,' he huffed in 1935. Einstein favoured an alternative idea of an underlying but as-yet inaccessible layer of reality containing hidden influences that 'told' the photon about the nature of the experiment to be performed on it, changing its behaviour accordingly.

There is more to this than wild conspiracy theory. Imagine an explosion that sends two pieces of shrapnel in opposite directions. The explosion obeys the law of conservation of momentum, and so the mass and velocity of the pieces are correlated. But if you know nothing of momentum conservation, you could easily think that measuring the properties of one fragment determines the properties of the other, rather than both being set at the point of explosion. Was a similar hidden reality responsible for goings-on in the quantum world?

In 1978 John Archibald Wheeler, one of the twentieth century's leading theoretical physicists, lit on a very strange idea to test

FIGURE 7.2 Wave–particle duality is the central mystery of quantum mechanics.

this. Its aim was to settle the issue of what told the photon how to behave, using an updated version of the double-slit experiment. Photons would be given a choice of two paths to travel in a device known as an interferometer. At the far end of the interferometer, the two paths would either be recombined or not. If the photons were measured without this recombination – an 'open' interferometer – that was the equivalent of putting a detector at one or other of the slits. You would expect to see single particles travelling down one path or the other, all things being equal, splitting 50:50 between the two (see Figure 7.3).

Alternatively, the photons could be measured after recombination – a 'closed' setting. In this case, what you expect to see depends on the lengths of the two paths through the interferometer. If both are exactly the same length, the peaks of the waves arrive at the same time at one of the detectors and interfere constructively there: 100 per cent of the hits appear on that detector and none on the other. By altering one path length, however, you can bring the wave fronts out of sync and vary the interference at the first detector from completely constructive to totally destructive, so that it receives no hits. This is equivalent to scanning across from a bright fringe to a dark one on the interference screen of the double-slit experiment.

Both paths

Wheeler's twist to the experiment was to delay choosing how to measure the photon – whether in an open or a closed setting – until after it had entered the interferometer. That way, the photon could not possibly 'know' whether to take one or both paths, and so whether it was supposed to act as a particle or a wave.

It was almost three decades before the experiment could actually be done. But the result was worth the wait. Whenever

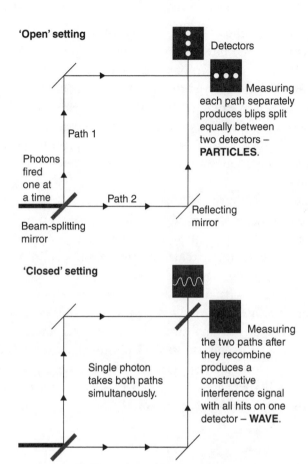

FIGURE 7.3 According to how it is set up, an interferometer can be used to 'prove' that light is particles, waves or neither.

the researchers chose at the last instant to measure the photons with a closed interferometer, they saw wave interference. Whenever they chose an open interferometer, they saw particles. Then, things got even stranger. In December 2011 another team proposed extending Wheeler's thought experiment. Their new twist was that the decision of how to measure the photon, as a particle or as a wave, should itself be a quantum-mechanical one – not a definite yes or no, but an indeterminate, fuzzy yes-and-no.

There is a way to do that: you use light to control the detector designed to probe the light. First you prepare a 'control' photon in a quantum superposition of two states. One of these states switches the interferometer to an open, particle-measuring state, and the other to a closed, wave-measuring state. Crucially, you only measure the state of the control photon after you have measured the experimental 'system' photon passing through the interferometer. As far as you are concerned, the system photon is passing through an interferometer that is both open and closed; you do not know whether you are setting out to measure wave or particle behaviour. So what do you measure?

Shades of grey

The results were unnerving: what you see depends on the control photon. If you look at the measurements of the system photons without ever checking the corresponding measurements of the control photons – so never knowing what measurement you made – you see a distribution of hits on the two detectors that is the signature of neither particles nor waves but some ambiguous mixture of the two. If particle is black and wave is white, this is some shade of grey.

Do the same, but this time looking at the control photon measurements as well, and it is like putting on a pair of magic specs. Grey separates clearly into black and white. You can pick out the system photons that passed through an open interferometer, and they are clearly particles. Those that passed through a closed interferometer look like waves. The photons reveal their colours in accordance with the kind of measurement the control photon said you made.

It gets yet stranger. Quantum mechanics allows you to put the control photon not just in an equal mix of two states, but in varying proportions. That is equivalent to an interferometer setting that is, say, open 70 per cent of the time and closed 30 per cent of the time. If we measure a bunch of system photons in this configuration, and look at the data before putting on our magic specs, we see an ambiguous signature once again – but this time its shade of grey has shifted closer to particle black than wave white. Put on the specs, though, and we see system photons 70 per cent of which have seemingly – but clearly – behaved as particles, while the remaining 30 per cent acted as waves.

What does this mean for our understanding of reality? In one sense, the results leave Bohr's side of the argument about quantum reality stronger. There is a tight correlation between the state of the control photon, representing the nature of the measurement, and the system photon, representing the state of reality. Make for more of a particle measurement, and you'll measure something more like a particle, and vice versa. As in earlier experiments, a hidden-reality theory à la Einstein cannot explain the results.

The outcomes of the latest experiments simply bear that out. 'Particle' and 'wave' are concepts we latch on to because they seem to correspond to guises of matter in our familiar, classical world. But attempting to describe true quantum reality with these or any other black-or-white concepts is an enterprise doomed to failure.

Could quantum weirdness hide an orderly reality after all?

Often brushed aside like a forgotten stepchild, a theory of quantum mechanics devised more than half a century ago may now share the stage with its better-regarded siblings. If it holds up, it might lend support to ideas that the universe is improbably interconnected across vast distances.

In 1952 the physicist David Bohm suggested that the quantum world only appears weird because we do not know enough about its underlying reality. Beneath the quantum weirdness, he said, reality is orderly, with all the particles in nature possessing definite positions and following definite trajectories.

Many recent experiments have suggested that no such hidden reality exists. However, they have only ruled out a specific class of theories in which the hidden reality of any particle is local, and not influenced by something far away. Bohm's ideas involve non-local hidden reality, in which everything depends on everything. In his universe, something happening in a distant galaxy is influencing you right now and vice versa, however minor the effect. Bohmian mechanics remains controversial but, after recent experiments have started to bear out its predictions, people might now be taking it a little more seriously.

The end of uncertainty

Heisenberg's uncertainty principle prevents us from knowing all there is to know about a system. But quantum entanglement – by connecting the states of distant objects, meaning that if we determine one, we determine the other – seems to give us a workaround. Are these two pillars of quantum theory fundamentally incompatible?

The now-favoured version of Heisenberg's uncertainty principle was constructed in 1988 by two Dutch physicists, Hans Maassen and Jos Uffink, using concepts from the theory of information devised by the American mathematician Claude Shannon and others in the years following the Second World War.

Shannon had shown how a quantity that he termed entropy, by analogy with the measure of thermodynamic disorder, provided a reliable indicator of the unpredictability of information, and so quite generally of uncertainty. For example, the outcome of the most recent in a series of coin tosses has maximal Shannon entropy, as it tells you nothing about the result of the next toss. Information expressed in a natural language such as English, on the other hand, has low entropy, because a series of words provides clues to what will follow.

Translating this insight to the quantum world, Maassen and Uffink showed how it is impossible to reduce the Shannon entropy associated with any measurable quantum quantity to zero, and that the more you squeeze the entropy of one variable, the more the entropy of the other increases. Information that a quantum system gives with one hand, it takes away with the other.

Quantum entanglement can have a distinctly weird effect on uncertainty. Suppose an observer named Bob creates a pair of particles, such as photons of light, whose quantum states are somehow entangled. Bob sends one of these entangled photons to a second observer, Alice, and keeps the other close by him in a quantum memory bank – a suitable length of optical fibre, say. Alice then randomly measures one of a pair of complementary variables associated with the photon: in this case, polarizations in two different directions. Her measurement will be governed by the usual rules of quantum uncertainty and can only ever be accurate to within a certain limit. In Maassen and Uffink's terms, its entropy will be non-zero. Alice tells Bob which of the quantities she measured, but not the value that she obtained.

Now comes the central claim. Bob's job is to find out the result of Alice's measurement as accurately as possible. That is quite easy: he just needs to raid his quantum memory bank. If the two photons are perfectly entangled, he need only know which quantity Alice measured and measure it in his own photon to give him perfect knowledge of the value of Alice's measurement – better even than Alice can know it. Over the course of a series of measurements, he can even squeeze its associated entropy to zero.

Thought experiment

To many, this thought experiment is reminiscent of the famous 'EPR' thought experiment devised in 1935 by Einstein and his colleagues Boris Podolsky and Nathan Rosen. It, too, came to a similar conclusion: that entanglement could remove all uncertainty from one measurement, but not from both at once. In keeping with Einstein's general scepticism about quantum weirdness, he interpreted the tension between the two principles as indicating that quantum mechanics was incomplete, and that a hidden reality lying beneath the quantum world was determining the outcome of the experiments.

While that debate is now largely considered settled, the latest work opens up an entirely new perspective. Traditionally, debates about the validity of the uncertainty principle and the interpretation of the EPR experiment have remained distinct. Now there is another possibility: not that uncertainty is dead, but that uncertainty and entanglement are two sides of the same coin.

Where two particles are perfectly entangled, spooky action at a distance is in control, and uncertainty is a less stringent principle than had been assumed. But where there is no entanglement, uncertainty reverts to the Maassen–Uffink relation. This allows us to say how much we can know for a sliding scale of situations in between, where entanglement is present but less than perfect. That is highly relevant for quantum cryptography, the quantum technology closest to real-world application, which relies on the sharing of perfectly entangled particles. The relation means that there is an easier way to test when that entanglement has been disturbed, for example by unwanted eavesdroppers, simply by monitoring measurement uncertainty.

As for the duel between uncertainty and entanglement, it ends in a draw, with the two principles now becoming part of the same mathematical scheme. But, while that is true within the confines of quantum theory, we might be able to tell which is the stronger principle by zooming out and considering a mathematical framework more general than that of quantum theory (see 'Reality check', below).

Reality check

Our understanding of the quantum world appears to be fundamentally flawed. If we take quantum theory at face value, either relativity, causality, free will or reality itself must be an illusion. But which is it?

Reconciling entanglement with 'normal' physics is no easy task. Einstein believed that there must be some undetected influence that flies between the two photons. But whatever form that influence might take – a photon, some other exchanged particle or perhaps a type of wave – a good guess is that it will not travel faster than the speed of light. Thanks to Einstein's relativity, that is always seen as a kind of fundamental speed limit to any kind of usable information flying through the universe. Having that limit prevents all sorts of unpleasant consequences. Any faster-than-light channel might also be open to hijacking for nefarious purposes: you could use it to transmit information backwards in time. Allow violations of relativistic causality, and we could all be lottery millionaires.

Hidden physical influences of the less outlandish sort, which obey relativity, can be tested for relatively easily. First you separate two entangled photons by a huge distance. The second photon is sent away – to the International Space Station (ISS), say – with instructions to carry out a measurement at a precise time. An instant before that measurement occurs, you measure the first photon. Time it right, and there is not enough time for any influence to travel between the two, even at the speed of light.

Nobody has yet done the ISS test, but we have done similar things many times on Earth. Each time, when the report of the second measurement comes back, the weird influence has still

FIGURE 7.4 Can entanglement be reconciled with 'normal' physics?

been felt. The second photon responds to measurements as if it were aware of what happened to the first.

This leaves us forced to conjure up a reality outside of space-time – unless there is something fundamental that we have wrong. Violations of relativity are frowned upon because they violate our ideas of causality. We humans are easily seduced by the idea of causal order, looking back in time to trace the cause of any event. Even more basically, we are determined determinists, blithely assuming that every event actually has a cause. That seems to work reasonably well in our large-scale everyday world, but when it comes to the nitty-gritty of the underlying quantum reality, can we be so sure?

Causal laws

Theorist Časlav Brukner and colleagues at the University of Vienna in Austria set out to investigate whether quantum systems are subject, in theory, to the same causal laws as the rest of us. They started off from the classic situation in which two independent observers, Alice and Bob, make a measurement on a photon. The twist Brukner and his team added was quantum uncertainty, a principle that fundamentally constrains the amount of information you can extract from a quantum system – including information about time.

Brukner describes the scenario they uncovered as akin to having Alice walk into a room and find a message written by Bob. She erases it and writes a reply – then Bob comes in to write the original message that Alice has just replied to. In effect, just as quantum particles can be in two or more places at once, so seemingly can this particle be in two or more moments at once. The system can be simultaneously in the states 'Alice came into the room before Bob' and 'Bob came into the room before Alice'. And we cannot say whether Alice's measurement is ahead of Bob's measurement, or the other way round. But testing the results of these theoretical calculations in experiment will not be easy. Given the delicate nature of quantum states, any attempt to measure a quantum-mechanical superposition of causal orders destroys that superposition, collapsing it into a definite causal order.

The conclusion appears to be clear: causal order is not a fundamental property of nature. Causality is restored only when the parameters of the experiment are tweaked to make the particle behave more like familiar, classical particles. If we accept quantum theory as the most fundamental description of reality that we have, it means that space-time itself is not fundamental, but emerges from a deeper, currently inscrutable quantum reality.

But could quantum theory itself be the problem? For all its successes, perhaps all that randomness, uncertainty and spooky influence is just because quantum mechanics is incomplete. As currently formulated, at least, it might simply not supply all the information we need to explain why things are as they are. An analogy might be made with the laws of thermodynamics. They provide a foolproof, high-level description of how things work – heat always passes from the hotter to the cooler – while saying nothing about the underlying dynamics of individual atoms that makes that happen.

To investigate this possibility, researchers have taken a look at what would happen in those classic Alice-and-Bob-type experiments if an underlying theory were to provide an additional, arbitrary amount of information about the correlations between two entangled particles. Do the outcomes of the measurements look any less random and unpredictable?

Mysterious unpredictability

The short answer is no. In any situation where both Alice and Bob can independently choose the type of measurement they make on their particle, additional information does not make their predictions of what will happen in experiments any more accurate than if they use quantum theory. The mysterious unpredictability of quantum mechanics has nothing to do with incomplete information, it seems.

Deep down, the universe is spontaneous. Fundamentally, there is no reason why a quantum particle has the properties it does: there is no hidden influence, no cast-iron cause and effect, no missing information. Things are as they are; there is no explanation (see Figure 7.5).

Some people find this so depressing that it leads them to question an even more fundamental assumption about reality and our relation to it. It lies in a little clause in the way most

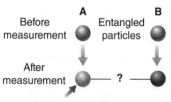

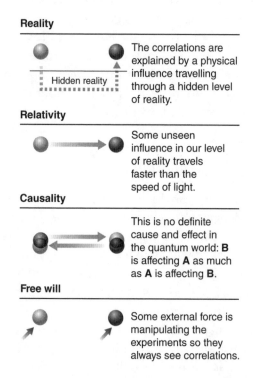

FIGURE 7.5 The strange correlations between quantum objects, however far apart they are, can be explained only by abandoning some fundamental assumptions.

investigations of quantum reality and quantum measurements are set up. To measure the value of some quantum phenomenon, you must first choose something to measure it relative to – the lab, the wind, the fly on the ceiling. Your choice influences the outcome of the measurement. But what if it is not actually your choice? What if something else were forcing your hand, making you perform the experiments such that the correlations always appear?

This takes us into the domain of human free will, a slippery territory where philosophers are usually more abundant than physicists. It sounds vaguely loopy, yet some serious physicists think that a lack of free will – that we are participants in something of a cosmic puppet show – might be the best way to save us from all the weirdness and loss of relativity and causality implied by quantum correlations.

The trouble is, nature has no interest in how we would like things to be. Put simply, our conceptions of reality, relativity, causality, free will and space and time can't all be right. But which ones are wrong?

The path to enlightenment

Obtaining a solid theoretical foundation for quantum theory has eluded scientists for more than a century. But six basic principles might be all it takes to make sense of it – and lead to a theory of everything.

We have become accustomed to the universe blowing our minds – perhaps too accustomed. So it is tempting to throw up our hands and say human brains can never grasp it. After all, the neat equations of quantum theory were developed not to encapsulate some universal principle, but in an ad hoc way to 'explain' weird experimental results. Quantum objects are described by wave functions that might or might not correspond to anything physical, and which exist in an abstract, multidimensional domain called Hilbert space. What is more, they evolve according to abstruse rules laid down in the Schrödinger equation. Erwin Schrödinger came upon this formulation haphazardly in 1925 by studying the equations of classical optics, which dealt with waves rather than particles. It works well, but it is not quite clear why.

It seems that the bedrock quantum mechanics rests on is made of information. Many theorists are coming to the conclusion that physical interactions can all be described as a form of information processing. Atoms carry information in their momentum, for instance: when two atoms collide like balls on a billiard table, their momenta change in a way that is equivalent to the way binary digits change as they go through a computer's logic gates. Rules governing information manipulation could ultimately determine what does and doesn't happen in our universe (see 'The fivefold way' below).

The fivefold way

Giacomo Mauro D'Ariano of the University of Pavia in Italy and his colleagues Giulio Chiribella and Paolo Perinotti have come up with five fundamental principles that must apply to any physical system to make a sensible measurement of it – as well as a sixth that they claim explains the weirdness of quantum measurements (see 'The path to enlightenment'). The five principles are as follows:

1 **Causality** – Stuff in the future cannot affect a measurement you are making right now.

2 **Distinguishability** – If a state is not too noisy, then there exists another state that can be distinguished from it.

3 **Composition** – If you know everything it is possible to know about all the stages of a process, then you know everything you can about the whole process.

4 **Compression** – There are ways to efficiently transmit all the information relevant to a measurement of a physical system without having to transmit the system itself.

5 **Tomography** – When you have a system with several parts, the statistics of measurements carried out on the parts is enough to identify the state of the whole system.

These basic principles might be all it takes to make sense of how the world works – and lead us towards a theory of everything.

Quantum vs classical

Several phenomena distinguish quantum from everyday, classical physics. One is superposition, where particles seem to be in two places at once, or spin clockwise and anticlockwise at the same time. Another is entanglement: two particles are entangled when measuring the properties of one instantly influences the properties of the other, no matter how far apart they are. Then there is the apparently random outcome of any measurement on a quantum system. You cannot say what the result of any one measurement will be before you perform it; you can only calculate the probabilities of different outcomes.

In the field of quantum information theory, an important distinction exists between different types of quantum state. For applications such as cryptography, some states are described as 'pure' – meaning that we know everything we can about such states. It is fairly straightforward, for example, to know every-thing about a single isolated hydrogen atom in its lowest energy state: this is a pure state. Then there are 'mixed' states, for which our information is incomplete. One particle of an entangled pair is in a mixed state: you cannot know everything about one member of the pair without taking its counterpart into account.

But the intriguing point is that both entangled particles together form a pure state. The two particles embody all the information there is to know about the quantum system. Could it be that all the messy uncertainty of the quantum world sim-ply stems from lack of information – making a measurement of a mixed state without having access to the larger pure state of which it is part?

If every mixed state were part of a pure state, it would be possible to describe each physical process with maximum information. Take the existence of such a perspective as a so-called purification principle, and then it is always possible for someone with access to the pure state to make measurements consistent with those of less well-positioned observers who are dealing with mixed states (see Figure 7.6). It does not make the weirdness go away, but it does make it explicable.

In general, caution is still necessary. For one thing, research-ers are still trying to develop a way to track the evolution of a quantum system. For that they need to work out where specific properties such as a particle's mass fit into their framework.

Focus too tight

Measure one particle's properties and the results may seem random. That's because it's really part of a wider system for which you don't have all the information.

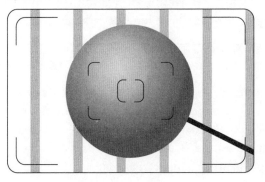

Proper perspective

Zoom out to measure properties associated with any linked particles and you find that measurements on the expanded system give you no surprises.

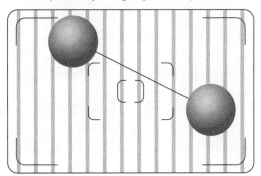

FIGURE 7.6 Understanding quantum weirdness may be a question of getting the perspective right.

If they can get that right, it might provide more than just a justification for quantum theory: it might be another route to combining general relativity with quantum mechanics. That could lead us to a description of quantum gravity, a 'theory of everything' researchers have long been keen to create.

Do wave functions need observers to collapse?

One alternative to explain the delicacy of superposition is that wave functions can collapse randomly, by themselves. Such objective collapse would be rare, but catching. Wait for a single particle's wave function to collapse and you could be waiting longer than the age of the universe. Group many particles together, however, and the chance swiftly escalates. With a few billion particles, you might have to wait only a few seconds for one wave function to collapse – and for that to set the rest off.

Such a proposal could explain quite a lot that is inexplicable about quantum theory. We never see ghostly quantum effects in large objects such as cats or the moon because, with so many interacting particles, their wave functions readily collapse or else never form. And in the early universe, it was only a matter of time before the wave functions of matter collapsed into an uneven distribution from which stars and galaxies could form.

Objective collapse theory has an intuitive explanation for the observer problem, too. The human body has upwards of one billion billion billion atoms, all of which contain yet more constituent particles. An observer meddling with even a carefully isolated quantum apparatus will inevitably become quantum-entangled with it, and their collapsed wave function then causes any uncollapsed wave function in the vicinity to collapse too.

Spontaneous wave function collapse makes stuff, too. When a wave function disappears, something new appears in its place – a definite position, a piece of information, a tick of energy. Each collapse can generate only a minuscule amount of energy, so we would not notice it on any everyday scale. But in the universe as a whole, this energy creation could be rather significant. It might even solve the biggest cosmological conundrum of all: the nature of the hitherto unexplained 'dark energy' that seems to be causing the universe's expansion to accelerate (see Chapter 8).

8

The quantum cosmos

To comprehend the birth and growth of our universe, we have relied for a century on Einstein's theory of gravity, general relativity. But it can only take us so far. Now the notions of quantum mechanics have begun to percolate out of their traditional home of particle physics to join relativity in the vast spaces between the stars. Although the true cosmic role of the quantum remains hazy – a matter of unproven theories and conjectures about wormholes, black holes and superstrings – many physicists think that these extraordinary new ideas foreshadow a unified theory of quantum gravity that will finally reveal the origins of the Big Bang and the nature of space and time.

The question of quantum gravity

We live in a quantum universe. Look closely, and you will find that we are all entangled waves of complex probability. In most areas of physics, this underlying quantum reality reveals an unforeseen richness to the world we live in. All the particles that we have discovered, together with the electromagnetic and nuclear forces, are described by quantum field theory, which can predict the results of every experiment we have ever performed. But there is one realm of physics where imposing the laws of quantum mechanics brings headache rather than joy: gravity.

Isaac Newton was the first to propose a law of physics that describes gravity. However, our modern understanding comes from Albert Einstein, whose theory of general relativity teaches us that what we call gravity is really the bending and warping of space and time. We cannot separate gravity from the space-time stage on which we live our lives. To reconcile quantum mechanics with gravity, we must work out how the laws of quantum mechanics, with their randomness and uncertainty, apply to space and time. This is the challenge of quantum gravity.

If you care only about what happens when the energies involved are small, then it is fairly straightforward to cobble together general relativity and quantum mechanics in a consistent manner, in what is called an effective field theory. But, at high energies, things get more complicated. Heisenberg's uncertainty principle tells us that the position of a particle is a little blurred – you can never say for sure where it is. In quantum gravity, these same ideas apply to space and time. The place where you are sitting is constantly fluctuating as it suffers quantum jitters. So, too, is the 'now' you are experiencing. Viewed over large enough distances and long enough times, we don't notice these fluctuations. This is where effective field theory works. But as we look at smaller scales, the randomness becomes stronger, with space and time suffering wilder fluctuations. The goal of quantum gravity is to make sense of these fluctuations on short distance scales.

Usually in science, we do an experiment to make progress, but this is difficult with quantum gravity. Our most powerful microscope is the LHC, a particle collider in CERN in Geneva. It explores the universe at a distance scale of 10^{-20} metres. Tiny as this is, it's still a million, billion times bigger than the scale where we expect to see the fluctuations of space and time. At the moment, our best experiments are nowhere close to seeing the effects of quantum gravity.

However, we are not restricted to performing experiments here on Earth. The universe is a big place and plays host to many extreme phenomena, so we could look for a situation where the effects of quantum gravity occur naturally. There are two places where we are sure that quantum gravity is important: at the point of the Big Bang and within black holes. So, to find out what happens when quantum mechanics meets the force of gravity, we may have to travel back to the first moment of the Big Bang or explore the interior of a black hole.

Black holes

A black hole is a region in space-time so warped that nothing can escape, not even light. According to general relativity, lying at the centre of a black hole is a point in time, known as a singularity, where the curvature of space becomes infinite. Now whenever an equation in physics gives you the answer infinity, you should really interpret this as a confession of ignorance: it is the mathematical way of saying 'I don't know'. In the case of black holes, general relativity is admitting that the meaning of the singularity can only be found elsewhere, in a quantum theory of gravity.

If we could observe what happens near this singularity, it would give us clues about how quantum gravity operates. Unfortunately, the singularity lies behind the event horizon of a black hole, a region from which light can never escape. If we want to explore it, we are obliged to jump over the event horizon.

Another singularity sits at the first moment of the Big Bang. Again, it is telling us that we simply do not know what happened. We don't know if time started with the Big Bang, or if there was something before. We don't even know if the meaning of time makes sense at the Big Bang. For these questions, we need a theory of quantum gravity. Once again, if we could see what was happening close to the Big Bang, then we might get some precious clues. But this too is hard. It is not just that the Big Bang happened a long time ago, but also that a rapid expansion thought to have occurred in the early universe, known as inflation, may have swept all effects of quantum gravity far from view.

There is something conspiratorial in all of this. Nature has ensured that we are shielded from the effects of quantum gravity. The mathematical physicist Roger Penrose formulated the 'cosmic censorship conjecture', which, roughly speaking, states that singularities are always hidden from view. This can be translated into a precise mathematical statement about the equations of general relativity. After 50 years of concerted effort by mathematicians and physicists alike, most think that the conjecture is true but no one knows how to prove it. Usually in mathematics, things that seem to be true but are hard to prove are pointing to some deep fact – but we do not know what deep lesson we can take from cosmic censorship.

There are other questions, many revolving around black holes, where we also are in desperate need of a quantum theory of gravity. These are insidious, since both general relativity and the cobbled-together effective field theory of quantum gravity happily give sensible answers. Except we now know that these theories are lying to us.

For example, in the 1970s Stephen Hawking (see Figure 8.1) taught us that black holes aren't quite black. Once you take the effects of quantum mechanics into account, they start to glow, emitting light and slowly evaporating before they eventually disappear. Hawking pointed out that this leads to a paradox.

FIGURE 8.1 Stephen Hawking at the Oxford Union, UK, in 2016

David Tong, on holding the world record for being wrong

David Tong is a professor of theoretical physics at the University of Cambridge, working on quantum field theory.

'We have photographs of the fireball that filled the universe during the Big Bang. We call this the cosmic microwave background radiation. The flickers in this fireball contain information about what was happening in the first few fractions of a second after the Big Bang and it's possible that, if we stare closely enough, we may see some hint of quantum gravity.

'About ten years ago, some friends and I suggested a scenario for what happened at this very early time. We used string theory, and ideas of extra dimensions and branes, to suggest a novel mechanism for inflation – the rapid expansion of the universe soon after the Big Bang. Our mechanism gave a very distinctive signature for the flickering seen in the Big Bang fireball.

'Then, five years ago, the Planck satellite took a photo of this fireball at the best resolution yet. And our predicted signal wasn't there. It's possible that the signal will still be found if we look more closely but, for now, our theory in its most naive form is simply wrong. It may seem odd to boast about this, except that our theoretical scenario took place 10^{-37} seconds after the Big Bang. It's astonishing that we can look back this far into the past and apply the scientific method, testing what did or didn't happen. In fact, we may hold the world record for being empirically wrong before anyone else was wrong. Being able to compare theory and experiment in such extreme conditions gives me some hope that we will eventually be able to figure out what happened during the Big Bang.'

Hawking's paradox

Take a book and burn it. Practically, you have lost the information that was written on the pages, but in principle that information is still encoded in the subtle correlations of air molecules or in the flickering light of the flames. The book may be gone but the information it carried lives on, albeit in a form that is hard to decipher.

Suppose, however, that you take the same book and throw it into a black hole. If you wait long enough, the black hole will evaporate away. Hawking asked: where is the information now? He showed in a detailed calculation using general relativity that it has vanished. It cannot be found in the light emitted by the black hole. Yet one of the fundamental tenets of quantum mechanics says that information cannot be lost. The evaporation of black holes seems to contradict quantum mechanics. This is known as the information paradox.

Since the 1970s there have been many further attempts to understand what happens to information inside a black hole. Most scientists think that it does ultimately escape so that there is no contradiction with quantum mechanics. Yet no one has been able to point to the flaw in Hawking's calculation. It appears that general relativity is lying to us. But how should we change the theory to fix this? No one knows. For an in-depth analysis of the information paradox and its implications for a theory of quantum gravity, see 'Black hole firewall: trouble on the edge' below.

The universe may be giving us some other cryptic clues to quantum gravity. One of these is dark energy, our name for whatever is accelerating the expansion of the universe. Obviously, we would like to understand what it is. The good news is that the random quantum fluctuations of empty space could act like dark energy, causing the expansion of space to accelerate. Unfortunately, when we compute how fast this expansion must be, we get an answer that is too big by a factor of at least 10^{60}. This is one of the worst predictions in the history of science. Clearly, we are missing something important, something related to the way space reacts to quantum matter. But what is it? Again, no one knows.

The questions of quantum gravity are, at heart, the question of what space and time mean in a quantum universe. We have tried many approaches to quantum gravity. The best developed is string theory, which among other things provides a proof-of-principle that quantum mechanics and general relativity can happily coexist. There are a number of other ideas, but as yet none of these approaches can answer the basic questions that drove these developments in the first place. What happened at the Big Bang? What replaces the singularity in the black hole? How does information escape a black hole? These questions remain for future generations to solve.

Will we never be able to apply quantum mechanics to gravity because the universe does not operate on one coherent set of physical laws?

String theory shows that it is at least possible to reconcile quantum mechanics and gravity. But that does not mean that it is the way that nature chose. The laws governing our universe might be logically inconsistent, but there is some kind of conspiracy in the universe preventing us from ever exposing that inconsistency in an experiment – like a physical manifestation of Kurt Gödel's incompleteness theorem in mathematics. The fact that quantum gravity effects occur in places where we cannot observe might be taken as evidence of this conspiracy.

Rethinking how relativity meets quantum physics

One rules on the atomic scale; the other rules across the cosmos. If physicists can meld quantum theory and relativity together, they hope to create a theory of everything that shows how the whole universe works at a fundamental level.

So far, attention has focused on what happens in high-energy conditions that existed in the earliest moments of the Big Bang, where both theories should offer answers. The problem is that experimenting with such theories is incredibly difficult. You have to build an accelerator the size of the solar system, says Roger Penrose at the University of Oxford.

But perhaps the quantum world has more in common with relativity than we thought. According to Penrose, we have been doing experiments for decades that combine quantum theory and gravity, and with a few tweaks they might offer a different route to the revelations we seek.

Until now, the interplay between some oddities of physics has largely been ignored. Take the fact that atoms and small molecules can exist in two places at once, the phenomenon known as superposition (see Chapter 2). General relativity says that mass distorts space and time, so, in superposition, is an atom's mass creating two distinct distortions in space-time, and thus exerting a gravitational pull on itself (see Figure 8.2)? More fundamentally, it is questionable whether relativity even allows superposition. The reason quantum reality is so different from our everyday experiences could be right under our feet.

A new generation of experiments could answer these questions by probing how gravity affects delicate quantum states. A classic way to see the effect of superposition is to fire an atom at a screen with two slits. The result is an interference pattern that forms in a detector placed behind the slits:

Classical object deforms space-time to create
a gravitational field.

Would a quantum particle existing in two places at
once create two gravitational fields? If so,
where would they be?

FIGURE 8.2 Weighty matters: all objects leave their mark in space-time, but how would a particle that can be in more than one place at the same time affect it?

a series of well-defined patches where the atoms appear to hit the detector, alternated with blank spaces where no atom seems to land. The only explanation for such a pattern is that the atom goes through both slits, and the two parts of its wave function interfere before it reaches the detector. If you then add another detector to see which slit the atom went through, it destroys the interference pattern.

There are many ideas for why such a thing might happen. Most are to do with information loss: reading the atom's path forces the atom to choose one path or the other and prevents it taking both. Experiments have shown that there does not even have to be a detector: heating the atom up, so that it emits thermal photons that could be used to infer its position, seems to be enough to weaken the interference pattern.

For larger objects, superposition seems much harder to attain. We have made interference patterns with molecules composed of hundreds of atoms, but the more massive they get, the shorter-lived the superposition. This could be connected with information loss, but others suspect another influence at work. They suggest that gravity is the real reason why massive collections of atoms, including us, do not behave like quantum particles.

Gravity's role

Testing this idea is far from easy because superpositions of atoms are such delicate things. But our ability to protect them from heat, vibrations and other disturbances has been improving very quickly, which means that we can start to get to get to grips with gravity's role.

For instance, Cisco Gooding, who earned his PhD at the University of British Columbia (UBC) in Vancouver, Canada, and his former supervisor at UBC, Bill Unruh, are looking at how an atom in a superposition experiences time as it flies along different paths and then recombines to produce an interference pattern. An atom can be thought of as a tiny oscillator, a bit like a clock's pendulum. Send it along two different space-time paths and it becomes two clocks ticking differently; when they come back together, those two clocks don't necessarily agree, says Gooding. That should be enough to degrade the interference pattern in predictable and detectable ways.

Igor Pikovski at Harvard University has another plan based on time anomalies. Working with Časlav Brukner's group at the University of Vienna in Austria, they think we could put a clock in a superposition of two different heights from the ground. That would mean that the two parts of the superposition exist in different parts of the Earth's gravitational field.

According to general relativity, clocks run faster in a weaker gravitational field. Over your lifetime, your head ages 300 nanoseconds more than your feet.

For a one-atom clock in a superposition, this creates a problem. The fact that the atom records different time in different places gives away information on the atom's position, and this destroys the coherence. As the two times diverge, the atom would be forced back to being at one height or the other. In other words, time dilation due to gravity can explain why we do not see quantum superpositions in our everyday world. This can be tested using 'atomic fountain' techniques that push atoms upwards through microwave fields to create ultra-accurate interferometers.

Other experiments involve a different kind of superposition. Dirk Bouwmeester at the University of California, Santa Barbara (UCSB), and Markus Aspelmeyer at the University of Vienna are independently making mirrored cantilevers. These structures look rather like springboards that exist in two configurations at once. When a photon in superposition hits the mirror, it can put the cantilever into a superposition of being both vibrating (as if a diver had just left the springboard) and undisturbed. This was first achieved a few years ago; and now Penrose has suggested that each part of the superposed springboard should create so much gravity for the other that they collapse back into one.

The challenge for Bouwmeester and Aspelmeyer's teams is to make the superpositions last long enough to investigate this effect. One of the problems with the diving boards is that it is hard to disconnect them from their environment. This results in superpositions collapsing because of vibrations transmitted through the apparatus, rather than gravity.

Ticking atoms

Making and studying superpositions of large objects – large in quantum terms, anyway – is new territory for researchers. And, not surprisingly, there are other ideas for why reality ceases to be quantum at larger scales.

One suggestion is that we need to revise quantum theory itself. A more elaborate version called Ghirardi–Rimini–Weber theory includes a phenomenon known as spontaneous localization, which makes superpositions impossible for objects composed of more than a certain number of particles. This theory suggests that the distribution of mass – its density – is what matters. We may find out that particular answer fairly soon. Markus Arndt's group at the University of Vienna has been repeating the double-slit interferometer experiment with ever-larger objects. Arndt believes that spontaneous localization would kick in with particles of a mass of somewhere between 100,000 and 100 million protons. They need to rule out spontaneous localization before pointing the finger at gravity.

People disagree about how much such work will illuminate the search for a theory of everything. Many believe that such a theory is as distant a prospect as ever. But if we find that gravity is interfering with the quantum world, that could be a good start. Gooding thinks that we could have answers within ten years. That is progress.

Black hole firewall: trouble on the edge

Paradoxes abound in physics. There's the cat that can be dead and alive at the same time; the time traveller who kills his own grandfather; the twins who disagree on their age after one returns from a near-light-speed trip to a neighbouring star. Each perplexing scenario has forced us to examine the fine detail of the problem, advancing our understanding of the theories behind it.

And now the paradoxes that cloak a black hole (see Figure 8.3) may help physicists finally build a theory of quantum gravity. Back in 1974, Stephen Hawking and Jacob Bekenstein of the Hebrew University in Jerusalem, Israel, showed that black holes radiate photons and other quantum particles in an agonizingly slow process that eventually causes the black hole to evaporate. Hawking spotted a problem with this picture. The radiation seemed so random that he surmised it could not carry any information about the stuff that had fallen in. So, as the black hole evaporates, any information that went in must eventually disappear. Yet this is in direct conflict with a central tenet of quantum physics, which says that information cannot be destroyed. The black hole information paradox was born.

FIGURE 8.3 An artist's impression of a black hole

Hawking thought that black holes destroyed information and the answer was to question quantum mechanics. Others disagreed. Hawking's idea came from his efforts to meld general relativity and quantum mechanics, a mathematical feat so elusive that he was forced to make approximations. American theoretical physicist John Preskill even made a bet with Hawking that black holes do not destroy information (see 'Hawking's change of heart' below).

Several arguments suggest that Hawking was wrong. One of the most compelling comes from thinking about what happens as the evaporating black hole gets smaller and smaller. If information cannot escape or be destroyed, then more and more has to be stored in an ever-shrinking volume. But, if this is the case, quantum theory says it should be very easy to make a tiny black hole when particles collide – we would see them being created at the Large Hadron Collider, for example, according to Don Marolf, a theorist at UCSB.

So if, instead, information is escaping from the black hole, perhaps Hawking radiation is not so featureless? In a simplified picture of Hawking radiation, the vacuum of space-time is constantly producing pairs of virtual particles, which pop into existence and quickly disappear. This changes near a black hole's event horizon, considered the point of no return for anything falling in. Occasionally, one of the pair is sucked into the black hole while the other escapes as Hawking radiation.

Heat and quantum information

If Hawking radiation is carrying out quantum information, then that creates a problem. Information and heat are linked, which should mean that the particles just inside the event horizon become immensely energetic as information is transferred to their partners outside – creating a wall of fire hot enough to burn up anything, or anyone, falling into the black hole.

The idea of firewalls seemed so preposterous that physicists started looking for other ways to transfer information out of a black hole. One possibility has been put forward by Steve Giddings, also at UCSB, who showed that if quantum theory breaks down in the vicinity of the event horizon, then it is possible to transfer information from within the black hole to distant outside regions, and thus avoid creating a firewall. But, for it to work, Giddings had to relax allow faster-than-light information transfer, which is forbidden by relativity.

Enter Joe Polchinski at UCSB. He reckoned he could crack the problem by combining Giddings's model with earlier work carried out by Leonard Susskind at Stanford University. Susskind put forward three postulates. The first was simply that information is not lost. The other two involve two observers called Alice and Bob who are approaching a black hole (to find out more about Alice and Bob in a different context, see Chapter 2). Intrepid Alice crosses the black hole event horizon. Cautious Bob stays outside. Susskind postulated that Bob sees nothing unusual as he sits outside the black hole, and that Alice also sees nothing amiss as she crosses the event horizon. That is because the event horizon is not a physical boundary; it is just an ordinary patch of vacuum in an ordinary patch of space-time that curves gently.

A 2012 paper by Polchinski, Marolf and others showed that these postulates cannot all be true simultaneously. If information is not lost, the firewall still exists and Alice ends up burning to a crisp. Here's how it works. Let's say particle A of Hawking radiation comes out early in the life of the hole. Quantum theory says that particle A is fully entangled with another Hawking radiation particle, B, that emerges later in the life of the black hole.

Now, particle B is one of a pair of particles, B and C, produced at the horizon, and C has fallen into the black hole. Space-time at the horizon is assumed to be nothing special, as general relativity dictates: just gentle gravity and low curvature, with no firewall. This demands that the virtual particles produced at the horizon be entangled with each other. So B must be entangled with C. But since early and late Hawking radiation must be entangled, B is also entangled with A.

The monogamy of entanglement

Unfortunately, this violates a cherished principle of quantum mechanics known as the monogamy of entanglement. Simply speaking, it says that particle B can be fully entangled with A or C, but not both. So something has to give. A firewall would burn up general relativity. Or maybe quantum mechanics is wrong, and information is not preserved.

Susskind remains sceptical of firewalls, but he has argued that they could mean that the singularity thought to lie at the centre of a black hole instead migrates to the horizon. If they do form, then space-time as we know it may terminate at the horizon. The paradox would also be resolved if there is something special about space-time near a black hole, so that information can be transferred faster than the speed of light as in Giddings work – although that would be another blow to relativity.

Or maybe, as Preskill points out, there is a fourth possibility: 'none of the above, something we haven't thought of'.

Hawking's change of heart

In 1997 John Preskill famously made his bet with Stephen Hawking that black holes do not destroy information. The work that settled the bet began in the same year, when the Argentine-American physicist Juan Maldacena used the mathematics of string theory to show that the theory of gravity inside a black hole is equivalent to a quantum theory operating on the black hole's surface.

It sounds esoteric, yet Maldacena's work is remarkable. While we do not yet know how to describe a black hole in its entirety, we do know how to work with quantum theory on the surface. It also means that quantum mechanics is valid at a black hole's surface and that as the hole evaporates it does not lose information. One caveat is that the type of space-time Maldacena studied is different from the space-time of our universe, but his result is so compelling that physicists are loath to quibble – and in 2004 it convinced Hawking to concede that black holes do not destroy information after all. He honoured the bet with an encyclopaedia of baseball, which Preskill likened to a black hole because it was heavy and it took effort to get information out of it.

Entangled universe: could wormholes hold the universe together?

A cryptic email from Juan Maldacena to fellow physicist Leonard Susskind in 2013 gives a clue about how to solve the paradoxes swirling around black holes – and so perhaps unifying quantum theory with general relativity. It contained a single equation: 'ER = EPR'. That brief equation promises to forge

a connection between two very different bits of physics put
forward by Albert Einstein (see Figure 8.4).

Einstein's general relativity has never failed an experimental test, yet we know it is missing something. The theory
describes space-time as a malleable yet smooth and featureless backdrop to reality. Even in the extreme case of black
holes, space-time is smooth. But in the 1970s, physicists
Jacob Bekenstein and Stephen Hawking derived a strange
result: black holes have a temperature, and hence a property called entropy. This takes us into the realms of quantum
theory, where everything comes in discrete chunks. Entropy
measures how many ways you can organize a system's various
constituents – the arrangement of atoms in a gas, for example. More possible configurations mean higher entropy. But

FIGURE 8.4 Albert Einstein in 1912

if a black hole is just smooth space-time, it should have no substructure, and thus no entropy. For many, this points to a hole in Einstein's theory.

Einstein levelled a similar charge at quantum theory. In 1935 a paper he wrote with Boris Podolsky and Nathan Rosen brought to light a property of the quantum world in which two particles could instantly influence each other, even if they were at opposite ends of the universe. In Einstein's view, this 'spooky action at a distance' – quantum entanglement, as it became known – was preposterous. And around black holes, where entanglement meets general relativity, it creates all those paradoxes that we are still unable to resolve.

A startling insight from Maldacena in 1997 gave new hope in resolving this problem and understanding how gravity and quantum mechanics might meet. He conjectured that equations describing gravity in some volume of space-time were just the same as a set of quantum equations describing the surface of that volume. If you could solve the surface equations, you could get a viable theory describing gravity inside. Other physicists found that this 'Maldacena duality' worked, even though they did not know why.

In 2001 Maldacena provided an intriguing example, going back to a paper written by Einstein, again with Rosen, in 1935. This one exposed another peculiarity of black holes. It showed how two black holes might be connected by a short cut through space-time, known as an Einstein–Rosen bridge – or a wormhole. Maldacena's duality showed that a wormhole would form only if the surfaces of the two black holes were quantum entangled. In 2009 theorist Mark Van Raamsdonk of the University of British Columbia in Vancouver, Canada, worked out what would happen if you were to slowly reduce the amount of entanglement between the black holes. The answer was rather like pulling at two ends of a piece of chewing gum. The wormhole becomes thinner until it breaks, and you have two unconnected bits of space-time (see Figure 8.5). Reverse the process – increase the entanglement – and the wormhole starts to form again.

It took a few more years to achieve understanding and for Maldacena to send that excited email: ER = EPR. ER referred to the paper Einstein wrote with Rosen introducing the concept of wormholes; EPR referred to the paper he

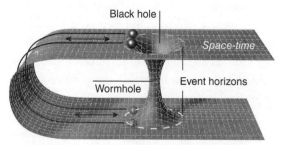

'Wormholes' connecting two black holes in different parts of space-time can exist – but only if particles on the black holes' surfaces are **quantum entangled**.

Break the entanglement, and the wormhole snaps too, suggesting that entanglement is the thread that binds space-time together.

FIGURE 8.5 The fabric of reality: it might be woven from quantum entanglement

wrote with Podolsky and Rosen the same year, introducing the concept of entanglement. What if wormholes and entanglement are two sides of the same coin: the same physics in different guises?

The principle provides some form of explanation for the phenomenon Van Raamsdonk's work had exposed, in which space-time in the form of wormholes could be created and destroyed simply by tweaking the amount of entanglement. This suggests a radical idea: that the whole of space-time is a manifestation of entanglement.

Does this mean that when quantum entanglement exists between two particles − say between photons in a lab experiment − they are connected by a microscopic wormhole? We do not know. So far, all of this work has been done with a space-time that is not expanding. Van Raamsdonk and others are working to extend the results to the sort of expanding, accelerating space-time that makes our cosmos.

Not everyone is convinced, but for those involved, this is the most positive lead yet towards a theory of quantum gravity that can unify the forces of nature. The ER = EPR principle is something that a theory of quantum gravity should obey, say Maldacena and Susskind. Susskind speculates further: quantum entanglement is a form of shared information, so space-time could be a manifestation of quantum information.

Given that Einstein developed the ideas of both wormholes and entanglement, one can only wonder what he would have made of it all.

Can quantum mechanics tell us what happened before the Big Bang?

General relativity says that the universe began with a singularity, with all its matter and energy compressed to a single point. It says that the laws of physics break down at a singularity, so it is impossible to predict what happens there.

But some cosmologists suggest that there may have been a 'big bounce', in which our universe rose from the ashes of an earlier cosmos that ended in a 'big crunch' – a process set to repeat when the current universe comes to an end. The big bounce has been modelled using loop quantum gravity, but this approach ran into trouble when physicists looked into the detail.

A new idea from Neil Turok of the Perimeter Institute in Waterloo, Canada, and Steffen Gielen at Imperial College London also gives us a bounce. They use a principle from particle physics: at very high energies, matter behaves like light. In particular, it becomes scale-invariant – the equations that describe its behaviour are the same, no matter the energy of the light or the size of universe that contains it. According to Turok, this means that the universe can shrink to zero and reappear, and the light is none the wiser.

Applying that principle to a universe that is completely smooth and the same in all directions, they predict a cosmos that bounces through the singularity in a process similar to quantum tunnelling, which allows electrons to pass through walls or other barriers. The next step is to drop some of those assumptions and try to make a universe that includes quantum fluctuations that give rise to large-scale structures such as galaxies.

Conclusion

Out of Max Planck's mathematical manoeuvre back in 1900 has grown an entirely new view of the world, where our reliable old reality is replaced by uncertainties, split identities and spooky connections that can stretch across the cosmos.

For more than a century, quantum mechanics has ruled the microworld of particles, but now – thanks to a new flowering of quantum technology – it is set to emerge even further into the macroworld, in the form of everyday consumer devices that tell us where we are and what the time is, reveal buried treasure, see round corners, and perform with ease what were once unfeasible feats of calculation.

Yet even as we learn to harness the full weirdness of quantum mechanics, we cannot agree on what the theory actually means. Does it point to a special status for consciousness, a skein of tramlines that guide particles, a multitude of universes? Does it threaten causality? Is there a more fundamental layer below the quantum? Our ideas about quantum mechanics seem to be in some kind of quantum superposition.

To collapse this philosophical wave function and find the true nature of quantum reality, some say we should look to the heavens. The answers may be written in the cosmic microwave background, or on the event horizon of a black hole. Or perhaps we just need a traditional scientific revolution: mix together a new generation of researchers with open minds, a little genius and luck and technology, and out comes some undreamed-of idea that really makes sense. Or maybe we can learn to live with the weirdness? Simply by playing around with quantum simulators and other technological toys that exploit superposition and entanglement, we might come to find these phenomena less alien and more intuitive, and wonder why our ancestors could not get out of their classically-minded rut.

Failing all that, maybe the answer awaits a giant quantum computer of the future, and someone working out the right question to ask it.

One hundred ideas

This section helps you to explore the subject in greater depth, with more than just the usual reading list.

Nine spots for quantum tourism

1 **Copenhagen**, Denmark: This city played an enormous role in the birth and development of quantum theory. Start the tour at Ved Stranden 14, where Niels Bohr was born on 7 October 1885. This was Bohr's grandparents' house and a plaque marks the building's famous resident. Move on to Bredgade 62, where Bohr grew up until he received his doctorate in 1911. This is now the Museum of Medical History. Also visit the Niels Bohr Institute, which was the meeting place of the brilliant young physicists who developed and refined quantum theory in the 1920s and 1930s. The institute lecture hall and Bohr's personal office have been kept intact and are open to visitors.
http://www.nbi.ku.dk/english/news/news13/niels-bohr-institute-named-historic-site/

2 **The Carlsberg Brewery,** Copenhagen: Niels Bohr received grants from the Carlsberg Foundation from 1911 and, after he won the Nobel Prize, the Carlsbergs gave him a house next to the brewery, the Carlsberg Honorary Residence. He moved in in 1931, and here he had many discussions with Einstein, Heisenberg and the like. The Carlsberg museum and exhibition centre is open to visitors and provides a glimpse of the house where Bohr lived until his death in 1962.
http://www.visitcarlsberg.com/

3 **Westminster Abbey**, London, UK: Here a stone is set in the floor commemorating Paul Dirac, inscribed also with his famous equation describing the quantum behaviour of the electron.

4 **Belfast,** Northern Ireland: John Stewart Bell, whose mathematical work in the 1960s paved the way for experimental tests of quantum weirdness, is arguably the greatest scientist to have come from Northern Ireland. His home town of Belfast named a street 'Bell's Theorem Crescent' in 2015 – to get round a rule on naming streets after people.

5 **CERN,** Geneva: This is not only where the theorist John Stewart Bell (see above) worked, but also where most of the breakthrough experimental discoveries were made that proved our model of matter based on quantum theory is right.

6 The island of **Heligoland,** off the north coast of Germany: This is where Werner Heisenberg stayed in 1925 and made his revolutionary formulation of quantum mechanics. A plaque to commemorate this breakthrough was unveiled on the island in 2000.

7 The **Hotel Metropole,** Brussels: The famous photograph of the pioneers of quantum mechanics was taken here (see Figure 1.5). This picture, which included Einstein, Bohr, Heisenberg, Schrödinger and other leading scientists of the day, was taken in 1927 during the Solvay Conference to discuss the new field of quantum mechanics.

8 **Ulm**, Germany, the birthplace of Albert Einstein: The house where Einstein was born is next to the railway station, and other monuments to the great man exist in this southern German city.

9 **Parallel universes**: At present there is no known way to visit parallel worlds (assuming they exist) even if

you could break the laws of physics. However, inter-universe travel might be possible if you are prepared to wait a (very very) long time. Physicist Michio Kaku of the City College of New York believes that, trillions of years from now, humans will have developed the technology to travel to other universes in order to escape the demise of our current one.

Thirteen anecdotes and strange facts

1 **Niels Bohr** was a respected Danish club soccer goal-keeper (his younger brother Harald played for the national team).

2 The French quantum pioneer **Louis de Broglie** initially did a degree in history and spent the First World War as a wireless engineer stationed at the Eiffel Tower.

3 **Niels Bohr's** house had free beer on tap.

4 **Erwin Schrödinger** introduced his cat not to illustrate quantum weirdness, but as a *reductio ad absurdum* to prove that quantum theory must be wrong.

5 The question of whether **Schrödinger** actually has a cat has not been resolved, but apparently when he was at Oxford University he had a cat named Milton. Milton's fate is unknown.

6 **Wolfgang Pauli**, who devised the exclusion principle named after him, was obsessed with the number 137 and, according to legend, died in room number 137 of the Red Cross Hospital in Zurich in December 1958.

7 **Pauli** was also notorious for his ability to make experiments and equipment self-destruct, break or fail, simply by being in the vicinity. This has even been dubbed the 'Pauli effect'.

8 **Schrödinger** had numerous affairs – which were alleged to inspire his eureka moments. When exiled to Dublin during the Second World War, he appeared with two 'wives', and fathered at least two daughters by different mistresses.

9   The often bow-tied **Schrödinger** was a great student of Eastern religions, and he also wrote poetry.

10 **Max Planck,** the founder of quantum theory, was deeply religious and in 1937 wrote: 'Both religion and science need for their activities the belief in God.'

11 **Ettore Majorana** was a quantum particle theorist whose life has echoes of quantum theory. He disappeared without trace on a boat journey from Palermo to Naples in 1938. His body was never found, but he was never conclusively sighted ever again, so he lived on in a Schrödinger-cat state, simultaneously alive and dead (although the Rome Attorney's Office did conclude in 2015 that between 1955 and 1959, at least, he was alive and living in Venezuela).

12 **Albert Einstein** was constantly stopped by people on the street, so devised a ruse. With a heavy accent he said: 'Pardon me, sorry! Always I am mistaken for Professor Einstein.'

13 **Paul Dirac** was a man of legendary taciturnity with a strange love of Mickey Mouse and, in his latter years, an obsession with the singer Cher.

Ten quotes

The pioneers of quantum mechanics were not entirely comfortable with the weirdness they discovered.

1 'Actually, I did not think much about it.' Max Planck, on his discovery that energy can only exist in certain amounts called quanta.

2 'Spukhafte Fernwirkung.' This phrase of Albert Einstein's means 'spooky action at a distance', which he used to deride the concept of entanglement.

3 'God does not play dice with the world.' Einstein used variants of this quote many times, and it is still the most quoted by anyone expressing their bafflement at how the seemingly solid world of classical reality can emerge from the fuzzy uncertainties of the quantum realm.

4 'Anyone who is not shocked by quantum theory has not understood it.' Niels Bohr.

5 'We are all agreed that your theory is crazy. The question that divides us is whether it is crazy enough to have a chance of being correct.' Bohr, to Wolfgang Pauli in 1958.

6 'Not only is the universe stranger than we think, it is stranger than we can think.' Werner Heisenberg's explanation for why we cannot quite accept what quantum theory is telling us.

7 'I don't like it and I'm sorry I ever had anything to do with it.' Erwin Schrödinger's response to his interpretation of quantum mechanics.

8 'I have done a terrible thing, I have postulated a particle that cannot be detected.' Wolfgang Pauli, after proposing the existence of the neutrino.

9 'Hell … doomed from the beginning.' Hugh Everett's journey to Copenhagen in 1959 to explain his many-worlds hypothesis to Niels Bohr did not go well.

10 'Shut up and calculate!' This infamous characterization of how many quantum physicists – especially those who adhere to the dominant 'Copenhagen interpretation' – deal with the philosophical conundrums raised by their subject is often attributed to Richard Feynman, who was not short of an aphorism or two. In fact, it seems first to have been used by physicist David Mermin in 1989, a year after Feynman's death – although, in the true spirit of quantum theory, not even Mermin is entirely certain.

Seven quantum jokes

1 A cop pulls Heisenberg and Schrödinger over for speeding. The cop asks Heisenberg, 'Do you know how fast you were going?' Heisenberg replies, 'No, but we know exactly where we are!' Cop gets annoyed and tells Heisenberg to open the boot of the car. 'Hey, did you know there's a cat in here that's dead?' the cop shouts. Schrödinger angrily replies, 'Well, he is now.'

2 Q: What did one photon say to the other photon?

   A: I'm sick of your interference.

3 Graffiti on wall: *Heisenberg might have been here.*

4 If the *Titanic* had struck a Heisenberg, would it still be floating?

5 Why are quantum physicists useless in bed? Because when they find the position, they can't find the momentum, and when they have the momentum, they can't find the position.

6 Why didn't the quantum particle cross the road? It was already on both sides.

7 Q: What's the difference between a car mechanic and a quantum mechanic?

   A: The quantum mechanic can get the car inside the garage without opening the door.

Two quantum limericks

1 'Einstein, Podolsky and Rosen' by David Halliday

Two photons, close-coupled at start,
Flew several parsecs apart.
Said one, in distress, 'What you're forced to express
Removes any choice on my part.'

2 'Fussy Electrons' by David Morin, Eric Zaslow, E'beth
Haley, John Golden and Nathan Salwen

An electron is sure hard to please.
When spread out, it sometimes will freeze.
Though agoraphobic, it's still claustrophobic,
And runs off when put in a squeeze.

Five names used to label ...

1 **Niels Bohr** has the element bohrium and a moon crater named after him.

2 **Albert Einstein** has the element einsteinium, the Bose–Einstein condensate (a state of matter), and a moon crater named after him – among other things.

3 **Max Planck,** the German physicist and founding father of quantum mechanics, also has a crater, on the far side of the moon, named after him, as well as the European Space Agency's Planck Space Observatory, which created the highest-resolution map yet of the cosmic microwave background.

4 **Enrico Fermi**, the Italian physicist who created the first nuclear reactor, has the element fermium named after him, as well as the elementary particles called fermions.

5 **Paul Dirac** has a particular type of fermion named after him: the Dirac fermion (technically, this is a fermion that is not its own antiparticle). The Fermi–Dirac condensate, a state of matter, derives its name from both him and Fermi.

Five important (or unlikely) correspondences

1 The letters written by **Niels Bohr** and **Werner Heisenberg** about their famous meeting in Copenhagen in September 1941 have been the subject of much historical interest. Did Heisenberg head to Copenhagen to discuss his moral objections to working on a German atomic bomb? It is not certain.

2 **Wolfgang Pauli** and the psychoanalyst **Carl Jung** had a long correspondence, some of which discussed their shared obsession with the number 137. Many of these letters are published in a book, *Atom and Archetype.*

3 **Pauli** was a close friend of **Bohr** and **Heisenberg**, with whom he exchanged many letters discussing his ideas. The letters he received are now archived at CERN.

4 In 1932 **Albert Einstein** wrote to the psychoanalyst **Sigmund Freud** about whether it was possible to curtail humanity's violent tendencies. Freud was pessimistic on this matter in his reply.

5 In the 15 May 1935 edition of the journal *Physical Review,* **Einstein** together with colleagues **Boris Podolsky** and **Nathan Rosen** famously published an article entitled 'Can Quantum-Mechanical Description of Physical Reality Be Considered Complete?' outlining their now famous EPR paradox. Bohr's response was published five months later in the same journal – with the same title. The debate between Bohr and Einstein on this matter is now legendary.

Five examples of 'fruitloopery'

New Scientist magazine's Feedback column of bizarre stories, implausible advertising claims and confusing instructions has proposed that such 'fruitloopery' is most readily detected by uncalled-for use of the word 'quantum'. Here are some prime examples:

1 **Quantum jumping:** This technique offers 'a universe of infinite possibilities'. It gives you the chance to put right what once went wrong in your own life, by tapping into the accumulated wisdom of the many alternative versions of yourself living in the multiverse.

2 **Quantum colour therapy:** 'In essence, we are all made of colour and frequency. As a result, whatever imbalances we might be experiencing on any level, whether it be physical, situational or emotional, can be translated into an imbalance in the colour/light frequencies we have radiating/missing within our quantum energy field.' Is there anything we can do about this? Yes, of course there is – we can give ourselves a dose of quantum colour therapy by buying a quantum balance crystal.

3 **Quantum clips:** These, apparently, are capable of manipulating certain inanimate material into a condition that mimics the quantum state of our living senses. (No, we have not got a clue what this means, either.)

4 **Quantum resource technology:** This handy innovation claims to make your home cleaner and more comfortable, by neutralizing the 'underlying disorderliness' caused by the 'random motion of electrons'.

5 **Quantum pendants** for pets: These use a special 'bio-energetic' process 'combining quantum physics, homeopathic principles and advanced computer software technology to work with your pets' unique energy to strengthen their immune system and create a frequency barrier that repels and controls fleas, ticks and mosquitos'.

*Five public information signs with instructions to exhibit
quantum behaviour*

1 'Queue both sides.'

2 'Strictly no parking on both sides of the road.'

3 'Use all doors to exit the train.'

4 'Please order your food from all till points.'

5 'Please ensure that BOTH doors are CLOSED when
entering the lift.'

Six ways to delve deeper into quantum computing

1 Try your hand at programming a quantum computer. IBM has put its old one online (http://www.research. ibm.com/quantum/) and the University of Bristol, UK, has put its quantum computer online too (https:// cnotmz.appspot.com/)

2 If you have at least $10 million to spare, you can buy your own D–Wave quantum computer. Google has one (though whether this computer is fully quantum is the subject of intense debate). D–Wave has also released an open-source quantum software tool, Qbsolv, to help boost the nascent field of quantum computer programming (https://github.com/dwavesystems/qbsolv) – with the caveat that access to a D–Wave system must be arranged separately.

3 Beforehand, you can bone up on the theory behind the machine by reading *The Fabric of Reality* by David Deutsch (1997).

4 *Quantum Computation and Quantum Information* by Michael Nielsen and Isaac Chuang (2000)

5 *Programming the Universe* by Seth Lloyd (2006)

6 *Decoding Reality* by Vlatko Vedral (2010)

Ten references to quantum mechanics and its inventors in music, movies, literature and theatre

1 **Schrödinger's cat** makes many appearances in literature. It was the title of a 1974 story by Ursula K. Le Guin, and the idea of it is explored by writers such as Terry Pratchett, Neil Gaiman and Douglas Adams. Then there's **Schrödinger's bat**, the title of the science book used by Lisa in *The Simpsons*.

2 Michael Frayn's award-winning play *Copenhagen*, first performed in 1998, is based around the famous 1941 meeting between **Niels Bohr** and **Werner Heisenberg** in the Danish capital.

3 Tom Stoppard's play *Hapgood*, first performed in 1988, uses quantum uncertainty/superposition as a structural device.

4 Mark Everett, son of **Hugh Everett**, deviser of the many-world interpretation of quantum mechanics, referenced his father in a 2005 song 'Things the Grandchildren Should Know':

> I never really understood what it
>   must have been like for him,
> Living inside his head,
> I feel like he's here with me now,
> Even though he's dead.

5 The 2008 James Bond movie *Quantum of Solace* had nothing to do with quantum mechanics. The title refers to author Ian Fleming's short story of the same name, which uses the term 'quantum' to mean 'minimum'.

The movie, however, is about Bond's quest to take down the shady Quantum organization.

6 In the TV show *Breaking Bad*, the lead character Walter White uses the name **Heisenberg** as his street name.

7 Doc Brown's loveable dog in the film *Back to the Future* was called **Einstein**.

8 *Quantum Jump* was a 1970s British band funk/jazz/rock band best known for their 1979 hit tune 'The Lone Ranger'.

9 Returning to more recent times, in 2016 Berlin-based singer Simone Jones released her quantum-influenced EP including a track titled **'Spooky Action'**.

10 In the US comedy series *The Big Bang Theory*, Howard and Leonard devise a **quantum gyroscope** to act as a guidance system – but then freak out when their invention attracts the attention of the military.

Nineteen items of recommended reading

1 Multiverse pioneer Hugh Everett's life was fascinating and tragic. Peter Byrne's book *The Many Worlds of Hugh Everett III* (2010) provides a detailed account.

2 For an excellent graphic exploration, see *Introducing Quantum Theory: A Graphic Guide to Science's Most Puzzling Discovery* by J. P. McEvoy (2007).

3 To find out about the burgeoning field of quantum biology, try *Life on the Edge: The Coming of Age of Quantum Biology* by Jim Al-Khalili and Johnjoe McFadden (2015).

4 *The Dancing Wu Li Masters* by Gary Zukav is the 2001 best-selling introduction to quantum physics.

5 *How the Hippies Saved Physics: Science, Counterculture and the Quantum Revival* by David Kaiser (2012) is an account of an eccentric group of physicists at Berkeley, California, who, in the 1970s, helped take physics in a new direction.

6 To round off the pothead reading selection, try *The Tao of Physics* by Fritjof Capra (1992).

7 *The Ghost in the Atom: A Discussion of the Mysteries of Quantum Physics* by P. C. W. Davies (2010) is an excellent backgrounder on the rival interpretations of quantum mechanics.

8 *Surely You're Joking, Mr Feynman!* by Ralph Leighton (1992) is a biography of Feynman, light on quantum but rich in bongos and other fun.

9 Delve into the complex life of Erwin Schrödinger in John Gribbin's *In Search of Schrödinger's Cat* (1985).

10 *Einstein – His Life and Universe* by Walter Isaacson (2008) is an excellent account of the great man.

11 And if you want your pet to learn about the quantum world, try Chad Orzel's *How to Teach Quantum Physics to Your Dog* (2010).

12 Another good introduction to the topic is *Quantum Theory Cannot Hurt You* by Marcus Chown (2014).

13 For an exploration of one of the quantum world's most eccentric characters, try Graham Farmelo's *The Strangest Man: The Hidden Life of Paul Dirac, Quantum Genius* (2010).

14 To find out more about Werner Heisenberg, try *Uncertainty* by David Cassidy (1991).

15 The Einstein–Bohr debate is outlined in *Quantum* by Manjit Kumar (2009).

16 Another book to tackle this debate is *Einstein, Bohr and the Quantum Dilemma* by Andrew Whitaker (1996).

17 On a deeper note, try *Quantum Mechanics and Experience* by David Albert (1994).

18 *Quantum Theory: Concepts and Methods* by Asher Peres (1995) is also an explanation of the meaning of quantum theory and the methods it uses.

19 *Quantum Mechanics: The Theoretical Minimum* by Leonard Susskind (2014) is a good introduction to the subject.

Four other ways to dig deeper

1 The Feynman lectures on physics:
   http://www.feynmanlectures.caltech.edu/

2 An archive documenting the life and work of Niels
   Bohr: http://www.nbarchive.dk/

3 Einstein online (www.einstein-online.info), a web
   portal from Germany's Max Planck Institute of
   Gravitational Physics (otherwise known as the Albert
   Einstein Institute), provides a wealth of information
   about the great man's theories and their applications.

4 An archive of Hugh Everett's work can be found here:
   http://ucispace.lib.uci.edu/handle/10575/1060

Glossary

Bell's inequalities Equations devised by physicist John Bell to test whether weird quantum influences over seemingly impossible distances could be caused by some other effect, like an unknown force or property.

Black hole A point of infinite curvature in space-time, around which the ordinary laws of physics break down – and quantum physics and Einstein's theory of relativity disagree. They mercilessly draw in matter and not even light can escape.

Classical physics Physics that predates quantum mechanics and relativity, such as Isaac Newton's laws of motion.

Copenhagen interpretation Although quantum mechanics has never failed any experimental test, physicists are undecided as to what it means for the nature of reality. This interpretation says that particles do not have definite properties until you measure them.

Decoherence A process whereby, once something gets large enough, it loses its quantum properties.

Dirac equation Devised by Paul Dirac, this married quantum physics and special relativity to describe an electron travelling at close to the speed of light.

Double-slit experiment A famous experiment that showed that quantum matter can behave as particles or waves, depending on whether it is being watched. In the experiment, photons of light are fired one at a time towards a screen with two slits. If the photons are watched, they will go through one of the slits and leave discrete blips on a screen. If the photons are not watched, an interference pattern forms instead, implying that the light is behaving as a wave and interfering with itself, passing through both slits simultaneously.

Entanglement The idea that, in the quantum world, objects can become linked, or entangled, such that changing one invariably affects the other, no matter how far apart they are.

General relativity Einstein's theory about how massive objects bend space and time around them, causing gravitational effects.

Heisenberg's uncertainty principle This states that, in the quantum world, objects do not have separate properties of momentum and position. They have a mixture of the two, which can never be completely separated.

Hidden variables As yet undiscovered forces or properties that some physicists believe could explain away quantum weirdness.

Many-worlds interpretation Another take on the implications of quantum mechanics (see Copenhagen interpretation). The many-worlds interpretation says that each time you make a measurement the universe splits, creating new realities. Each possible outcome of the measurement exists in one of these realities.

Measurement In a quantum context this does not have to be a deliberate action, but anything that gives away the properties of a quantum particle (and thus causes it to stop being in a superposition/causes it to adopt a definite state rather than existing in a combination of states).

Multiverse Our universe could be just one of a multitude of universes. There are different types of multiverse. The inflationary multiverse, for example, arises due to the exponential expansion of space-time, which means that far beyond the edge of the observable universe are countless other bubble universes, inaccessible to us. The many-worlds interpretation of quantum mechanics also involves the existence of countless universes, parallel to our own and interacting to generate quantum phenomena.

Non-locality The way quantum particles seem to be able to influence each other instantly over vast distances.

Observer People or things that perform a measurement.

Photons Individual particles of light.

Quantum computing Faster-than-regular computing based on quantum principles, which make it possible to run many calculations simultaneously.

Quantum cryptography Using the rules of quantum mechanics to protect information, so that it cannot be intercepted and read.

Quantum gravity Physicists are searching for a theory of quantum gravity to unite quantum physics, which describes very small systems, and general relativity, which describes gravity and the large-scale workings of the universe. At the moment, the two theories disagree about things such as what happens at the edge of a black hole.

Quantum information theory The quantum version of classical information theory, which is concerned with how information is stored and processed in different systems.

Quantum mechanics The laws explaining behaviour at the atomic and subatomic level, where particles move like waves, may be in several states at once and can have shared states connecting them across time and space.

Quantum teleportation A phenomenon in which the quantum states of one particle can be transferred to another, distant particle without anything physical travelling between them.

Quantum tunnelling Where a particle can travel through a seemingly impenetrable barrier using its wave properties.

Qubit Short for 'quantum bit'. In computing, a regular bit can be either a 0 or a 1. Qubits can be both at the same time.

Schrödinger's cat A famous thought experiment proposed by Erwin Schrödinder. A cat is put in a closed box where a deadly event can be triggered by an unpredictable event, like the decay of a radioactive particle. According to some interpretations of quantum mechanics, the cat is both alive and dead until you open the box to find out which.

Schrödinger equation This describes how a quantum system changes with time.

Special relativity Einstein's theory positing that the speed of light is always the same, irrespective of the motion of the person who measures it, and that space and time are deeply intertwined.

Spin (in a quantum context) A quantum property that many sorts of particle have, including electrons. You can think of it as the particle spinning on its axis, although this is not strictly the case.

Standard model of particle physics This covers the workings of three of the four forces of nature. It describes the interactions of force-carrying boson particles with matter-making fermions according to the mathematics of quantum field theory.

Superposition When a particle exists in more than one state simultaneously, like being in two places at once.

Uncertainty principle Heisenberg showed that if we know a particle's position precisely, we cannot be sure about its velocity, and vice versa – there is always uncertainty in one of the values. This is also true of other pairs of properties.

Wave-function collapse When a particle that is in a combination of states suddenly picks just one. This often happens when we make a measurement of a quantum system.

Wave–particle duality Tiny chunks of matter, like photons of light, can act like either particles or waves, depending on how you measure them.

Wormhole A short cut between two points in space-time. Some theories say that pairs of black holes could be connected by wormholes, thanks to quantum entanglement.

Picture credits

All images © *New Scientist* except for the following:

Figure 1.1: Hulton Archive/Getty Images

Figure 1.2: Universal History Archive/Universal Images Group/Rex/Shutterstock

Figure 1.3: Ullstein Bild via Getty Images

Figure 1.5: Science Source/Science Photo Library

Figure 2.2: Boyer/Viollet/Rex/Shutterstock

Figure 2.3: Bettmann/Getty

Figure 2.4: Bettmann/Getty

Figure 2.6: Axel Bueckert/Alamy Stock Photo

Figure 3.4: Ken Hively/*Los Angeles Times* via Getty Images

Figure 3.5: Roger Bacon/Reuters/Alamy Stock Photo

Figure 4.1: *Physics Today* Collection/Institute of Physics/Science Photo Library

Figure 4.4: Image courtesy of D-Wave

Figure 5.3: Volker Steger/Science Photo Library

Figure 6.1: ImageBROKER/Rex/Shutterstock

Figure 6.2: ImageBROKER/Rex/Shutterstock

Figure 6.3: PhotoAlto/Rex/Shutterstock

Figure 7.2: Joby Sessions/Future Publishing/Rex/Shutterstock

Figure 7.4: Siede Preis/Getty Images

Figure 8.1: Roger Askew/Oxford Union/Rex/Shutterstock

Figure 8.3: NASA/CXC/SAO

Figure 8.4: Rex/Shutterstock

Index

Note: Page numbers in italics refer to photographs.